Sustainable Development Goals Series

The **Sustainable Development Goals Series** is Springer Nature's inaugural cross-imprint book series that addresses and supports the United Nations' seventeen Sustainable Development Goals. The series fosters comprehensive research focused on these global targets and endeavours to address some of society's greatest grand challenges. The SDGs are inherently multidisciplinary, and they bring people working across different fields together and working towards a common goal. In this spirit, the Sustainable Development Goals series is the first at Springer Nature to publish books under both the Springer and Palgrave Macmillan imprints, bringing the strengths of our imprints together.

The Sustainable Development Goals Series is organized into eighteen subseries: one subseries based around each of the seventeen respective Sustainable Development Goals, and an eighteenth subseries, "Connecting the Goals," which serves as a home for volumes addressing multiple goals or studying the SDGs as a whole. Each subseries is guided by an expert Subseries Advisor with years or decades of experience studying and addressing core components of their respective Goal.

The SDG Series has a remit as broad as the SDGs themselves, and contributions are welcome from scientists, academics, policymakers, and researchers working in fields related to any of the seventeen goals. If you are interested in contributing a monograph or curated volume to the series, please contact the Publishers: Zachary Romano [Springer; zachary.romano@springer.com] and Rachael Ballard [Palgrave Macmillan; rachael.ballard@palgrave.com].

Koen Rens Wessels

Pedagogy of Entanglement

A Response to the Complex Societal
Challenges that Permeate our Lives

 Springer

Koen Rens Wessels
Utrecht University
Utrecht, The Netherlands

ISSN 2523-3084   ISSN 2523-3092   (electronic)
Sustainable Development Goals Series
ISBN 978-3-031-15789-9   ISBN 978-3-031-15787-5   (eBook)
https://doi.org/10.1007/978-3-031-15787-5

This Springer imprint is published by the registered company Springer Nature Switzerland AG
The registered company address is: Gewerbestrasse 11, 6330 Cham, Switzerland

Preface

The central theme in this book is the ontological move from individuality to relationality and the pedagogical implications this might have in a world permeated by complex societal challenges. The independent individual, autonomous in the sense of not needing anyone else and free in the sense of living just as one pleases, is a modern myth. Especially in Western countries such as the Netherlands, where I grew up, the strive to become such an individual has become deeply ingrained in all spheres of life. We are now facing the consequences. Such, at least, is my diagnosis, and I have tried to open up some hopeful pedagogical avenues in response. The book is structured as follows. I hope you will find it a stimulating read!

Chapter 1 – Toward a Pedagogical Response to Complex Societal Challenges

This chapter lays the foundation for the book, both in terms of aspirations and methodology. It positions the book as a pedagogical exercise in complexity thinking and develops its orientation toward *helpful perspectives* for situated pedagogical responses to complex societal challenges. Three interwoven methodological threads are introduced: (1) narrative inquiry with 12 teachers as co-researchers, (2) a pragmatic approach to literature study, and (3) autoethnography.

Chapter 2 – The Entangled Student

This chapter presents an initial inquiry aimed at arriving at a complexivistic understanding of the nature of students' relationship to complex societal challenges. Such an understanding is crucial for this book as it is this very relationship that the pedagogical response it seeks to develop is to be directed at. Through a critical reading of readiness- and space-dichotomies, the theoretical focus of the book becomes more clear as the notion of *entanglement* and its relational ontological underpinnings emerge and are grounded.

Chapter 3 – Diffractions

This chapter forms the empirical core of the book. It presents the majority of the inquiry with co-researchers – i.e., methods and outcomes – as it took shape around the premise of entanglement. Throughout the chapter, a diffractive approach to narrative research, based on a process of collaborative script-writing and rewriting, is developed.

Chapter 4 – On Relational Ontology and the Good Life

The outcomes of Chap. 3 form the input for Chaps. 4 and 5. In these chapters, helpful perspectives for a pedagogical response to complex societal challenges are developed. Herein, Chap. 4 focuses on *axiology* and Chap. 5 on *praxeology*. In doing so, the diffractive methodology introduced in Chap. 3 is further developed as an effort to bring the narratives and insights of Chap. 3 into conversation with complexivistic pedagogical literature. As such, Chaps. 4 and 5 position this book more thoroughly within a growing field of cutting-edge work connecting the theory of entanglement to pedagogical theory. In Chap. 4, the axiological perspectives explored and discussed are (1) *entanglement-orientedness*, (2) *entanglement-awareness*, and (3) *hopeful action.*

Chapter 5 – On Teaching the Entangled Student

This chapter continues the work done in Chap. 4, yet shifts focus to the pedagogical dimension of praxeology. The following perspectives are explored and discussed: (1) *inquiry within entangled phenomena*, (2) *practicing perceptiveness*, and (3) *practicing integrity.*

Chapter 6 – Living the Question of Integrity

This chapter portrays an autoethnographic account of my own practice of integrity throughout my inquiry with co-researchers. The aim of doing so is twofold: (1) to increase the "relatability" and usability of the six interpretations of integrity, and (2) to further enrich the book with the quality of transparency. The chapter concludes with the suggestion that engaging in the practice of integrity is not about "ticking boxes," but rather about nurturing an overall experience of meaningful effectiveness.

Chapter 7 – In Conclusion: Pedagogy of Entanglement

This chapter integrates the outcomes of the book into a heuristic that invites and helps teachers to explore and shape their situated pedagogical responses to the complex challenges that permeate contemporary society. In closing, this chapter highlight five openings to continue this quest, focusing on (1) educational research methodology, (2) teacher education, (3) schools as professional communities, (4) curriculum design, and (5) exemplary pedagogical practices.

Acknowledgments

In the spirit of the central themes of this book, I am of course far from independent myself. This book, therefore, can only be understood as a co-creation. In the entangled becoming of myself and this book, many have left their traces, and I am forever grateful. This awareness humbles me and moves me to embrace a dialogical way of being. For me, this book shall always be a celebration of those who co-produced it, and there are some that truly cannot stay unmentioned here. My academic mentors: Cok, Arjen, and George. The twelve teachers that wholeheartedly joined forces with me as co-researchers: Gerdien, Jonas, Simone, René, Aletta, Ineke, Petra, Attie, Joris, Kenneth, Arletta, and Sandor. My dear colleagues and friends from Utrecht University of Applied Sciences, Utrecht University, Wageningen University & Research, and the various groups for support and collaboration we were part of together. My pioneering friends from The Bildung Academy, and all the brilliant people who sparked me with hope and inspiration through our educational adventures. Those books and texts which captivated and transformed me, and became an invaluable voice in the chapters of this book. My loved ones, family and friends, human and canine. The cities of Amsterdam, Eindhoven, Istanbul, and Bodrum, and the buzz, character, and nature within and around. I thank you all for depending on you!

Contents

Toward a Pedagogical Response to Complex Societal Challenges

These turbulent times we live in, to put it mildly, do not particularly invite teachers to lean back, relax, and repeat the same curriculum year after year. For we live, indeed, in a world permeated by profound challenges and ongoing transformation (Harari 2016). As teachers, and consequently in educational research, we are inevitably confronted with the need to respond. We struggle, for instance, to respond to the increasingly flexible labor market (e.g. Scott 2017), the ecological crisis (e.g. Besley and Peters 2019; Lotz-Sisitka et al. 2015), the rise of fake news and post-truth politics (e.g. Ali-Khan and White 2019; Peters 2017; Zembylas 2020), the increasingly multicultural, globalized, and digital outlook of our societies (e.g. Banks 2008; Bond et al. 2018; Tatham-Fashanu 2021), the surge of the black lives matter movement (e.g. Dixson 2017), the rise of burnout among students (e.g. Walburg 2014), the impact of the COVID-19 pandemic on students' mental health (e.g. Son et al. 2020; Heikkilä and Mankki 2021), and, indeed the reality or looming threat of war (e.g. Hajir and Kester 2020).

The responsibility to navigate the challenges of our times can, of course, not be solely ascribed to teachers and educational institutions. Yet, as shall be my argument throughout this book, schools are not closed spaces separated from society but open places within society, and as such schools are inevitably complicit in the (re) shaping of our shared world. I believe, therefore, that an ambitious educational agenda is justified.

Following the United Nations' fourth Sustainable Development Goal (2015) – i.e. inclusive and equitable quality education for all – such an ambitious educational agenda has two crucial components. We need, firstly, to ensure that education is not a luxury for the rich and privileged, but self-evident and accessible for all, regardless of race, wealth, gender, location, age, and so forth. Secondly, education should be of a certain quality, so that it actually inspires humanity to navigate the challenges of our times with increasing sensitivity and wisdom (see, especially, target 4.7 of the Sustainable Development Goals). It is this second part of the United Nations' ambition that this book is primarily concerned with.

Each of the challenges articulated above, then, brings forward profound questions that deserve educational answers. For instance, if today's students are likely to become job-hoppers or, even, to grow into professions that do not yet exist, do we need a more trans-disciplinary education? And, do we perhaps need to emphasize such qualities as critical thinking, flexibility, and creativity? Can and should schools try to catalyze the transition to a more sustainable lifestyle to help solve the ecological crisis? How can we teach students to care for, and feel part of, nature? How, also, can education resist the attractions of fake news, post-truth politics, and polarizing forces, and teach students to critically evaluate their sources and their own reasoning? How can we strive for more inclusive and peace-oriented education, so that students

K. R. Wessels, *Pedagogy of Entanglement*, Sustainable Development Goals Series, https://doi.org/10.1007/978-3-031-15787-5_1

come to value and respect diversity? What kind of diversity are we actually talking about, and how inclusive are we to be exactly? What, furthermore, are the pros and cons of online education, and is, perhaps, a hybrid model ideal? And how far should we go in stimulating our students to have a healthy lifestyle (e.g. offline-online balance, rest-work balance, diet)? What is, actually, so challenging about growing up in today's society that many students struggle with mental problems (or are these struggles just inherent to growing up?), and are we, as educators, in some ways part of the problem?

Asking such and related questions is important – for each contemporary challenge deserves to be acknowledged in its own right and is unique in its particularities – and I encourage any project that focuses on any one of them. In this book, however, I follow another, complementary strategy, for my primary interest is not in what makes each challenge we face unique but in what they have in common, what binds them together. If we focus our inquiries on what it means to educate in a world-in-motion, a world permeated by profound transformations and challenges, we might end up with insights and strategies that can be utilized across contexts. Such generic insights and strategies are, in fact, not only important in light of the quality of education but can also contribute to the ambition to make quality education accessible for all; they can inspire and help teachers to shape meaningful educational processes within their unique contexts. Through such an approach, we can accomplish considerably more than if we start from scratch each time a new challenge confronts us or if we simply expect educational interventions to be equally effective across diverse contexts.

Confronted with Complexity

A promising angle for the kind of inquiry I thus hope to contribute to, is that of *complexity thinking*. Allow me to provide a brief introduction. In short, complexity thinking refers to 'a way of thinking and acting' that is 'concerned with the philosophical and pragmatic implications of assuming a complex universe' (Davis and Sumara

2006, p. 18). The idea of a "complex universe" was initially triggered by a growing difficulty to understand, let alone predict, newly discovered natural phenomena using the available laws of physics and mathematics. Complexity thinking started, consequently, as an approach dominated by natural scientists when various strands of so-called "complexity science" (e.g. cybernetics, systems theory, chaos theory) gained traction around the 1950s and 1960s (Alhadeff-Jones 2008; Davis and Sumara 2006). Today, however, complexity thinking is in no way limited to the natural scientist as this tendency and need to complexify has spread throughout the sciences, social sciences, and humanities. One of the most influential publications enabling this spreading is Rittel and Webber's paper introducing the notion of wicked problems (Rittel and Webber 1973). In their analysis, they explained several ways in which problems of social policy are of such a complex nature that 'the professionalized cognitive and occupational styles that were refined in the first half of this century, based in Newtonian mechanistic physics, are not readily adapted to contemporary conceptions of interacting open systems' (p. 156; n.b. "this century" refers to the twentieth century). By now, the accumulation of complexity research gives rise to a long list of ways in which the world we live in is complex rather than simple and, consequently, the problems we face wicked rather than tame (n.b. both Rittel and Webber, and Davis and Sumara provide quite extensive overviews). Without claiming to be exhaustive I would like to highlight three perspectives on complexity that I believe open up promising ways to think about education and pedagogy, and that particularly inspired me in the early stages of working on this book. In doing so, I will focus on the challenge of the ecological crisis for the sake of exemplarity.

Confronted with Multiple Perspectives and an Open Future

Decades of research, political debate, and social movements make it abundantly clear that it is an immense struggle to, in the face of the ecological crisis, reach one clear, shared narrative encompassing a singular problem-definition, vision for the future, and path to get there. Rather, society is permeated by a plurality of voices, stakes, and futures (Beckert 2016; Coyne 2005; Rittel and Webber 1973; Wahl 2016). As Rittel and Webber observe (1973, p. 163), when we plan for social change 'many parties are equally equipped, interested, and/or entitled to judge the solutions, although none has the power to set formal decision rules to determine correctness'. Indeed, we do not know for sure how best to arrive at an ecologically desirable human presence on Earth (e.g. which renewable energy source is most promising where and in what time-scale?, how are we to make the transition attractive and affordable for all?), nor what exactly such a desirable presence entails (e.g. what are the ethical limits to domesticating animals for consumption?). In our collective efforts to make sense of our predicament, in other words, the future is open (Crowell and Reid-Marr 2013; Wahl 2016). It is interesting to observe, therefore, that experimental curricula are being developed today around the idea of imagining new and diverse pathways into a more sustainable future (see, for instance, Hoffman et al. 2021). In complexity thinking, such an open approach to the future is crucial, for following its logic any attempt to predict and control the future is bound to be frustrated by contrasting voices and/or turns of events (Laloux 2014; Rittel and Webber 1973; Wahl 2016). Alternatively, as Wahl puts it, we need to practice the art of *living the questions*, in the hope that by allowing ourselves to not-know-yet, to imagine and experiment, and to be attentive to what emerges, we may live our ways into more regenerative cultures.

Confronted with Movements in Motion

Yet what renders any serious effort to find a way out of the ecological crisis complex is not only the simultaneous existence of a plurality of problem definitions, visions, stakes, and strategies, but also the fact that the very problems we are trying to resolve are continuously in motion (Davis and Sumara 2006; Wahl 2016). Average temperatures are rising, numerous species are on the verge of extinction, natural hazards are becoming more frequent and intense, innovative technologies are developing, ecological awareness is growing, and so forth. All our efforts to contribute to positive change are situated within these motions, are, to use a phrase borrowed from Akkerman et al. (2021), movements in motion. For instance, the move of planting seeds of diverse types of vegetation in a certain ecosystem to restore and strengthen bio-diversity is situated in a dynamic context of changes in soil composition, average temperature, and humidity in that particular ecosystem. If such a move has the desired effect, depends in a great deal not on the move itself, but on the way the motion it is situated in progresses under the influence of many other factors. For instance, if that particular region is confronted with extreme weather conditions soon after the seeds were planted (e.g. floods or extreme draught), all efforts may turn out to have been in vain. Whether such an extreme weather condition shall emerge, itself greatly depends on the ecological dynamics that transcend this particular local ecosystem, and so emerges a nested structure of local and increasingly global dynamics – of movements in motion – that co-specify each other (Davis and Sumara 2006; Laszlo 1996). Our actions, thus, are contextual and the context itself continually evolves. Such complexity demands, then, to nuance "what works" into – as Akkerman et al. (2021, p. 421) articulate it – 'what once worked for whom, when, where, how and for what purpose and with what kind of expansive possibilities'. Furthermore, the dynamics of the motions we move in prove to be rather hard, if not impossible, to predict and control (Wahl 2016). An important reason hereof, is that living systems demonstrate self-creativity (Laszlo 1996; Morrison 2008); they are able to evolve from within, to change their internal structure in response to changing conditions. For instance, if, for one reason or another, the population of a certain species in a certain region drops by 15 percent, a so-called cascade effect might move the local ecosystem into a state of accelerated change in which interactions between species are redefined and a new natural order emerges. Such emergent behavior, notably, tends not to present itself as a gradual, linear process, but rather occurs nonlinearly, as a sudden response to a certain accumulation of triggers (Crowell and Reid-Marr 2013; Davis and Sumara 2006).

Confronted with Interconnectedness

A consequence of the dynamic, nested nature of ecological processes, is that once we focus our attention on local ecosystems they appear to us as ambiguously bounded, meaning (Davis and Sumara 2006, p. 5): 'open in the sense that they continuously exchange matter and energy with their surroundings (and so judgments about their edges may require certain arbitrary impositions and necessary ignorances)'. Consequently, and so is the third and last perspective I wish to foreground here, we are confronted with interconnectedness, an important aspect of complexity thinking (Crowell and Reid-Marr 2013; Morrison 2008; Wahl 2016). An ecosystem, in this line of thinking, is an intricate web of relationality (Ingold 2008), in which myriad life forms co-constitute each other's existence. Yet, this interconnectedness and co-dependency present itself not only within one challenge – such as the challenge of finding a way out of the ecological crisis – but also between challenges, and in many ways wicked problems can be considered to be symptoms of other wicked problems (Rittel and Webber 1973). Consider, for instance, the following small narrative binding challenges of ecology, inequality, multi-culturalism, and mental well-being together (and I encourage you to imagine how the other challenges mentioned at the beginning of this chapter might be part of this narrative as well).

Imagine a farmer in Africa who becomes a climate refugee due to global warming. Humanity – and especially the rich with their high ecological footprints – is complicit in the farmer's predicament. This farmer, now, has to move to another district, or even another country, and thereby has his/her personal life heavily disturbed; the impact on this farmer's emotional well-being is likely to be large. Arriving in a new community, the farmer has to build a new life and thus contributes to the mingling of cultures and the challenges this poses.

Toward a Pedagogical Response

I began this book by highlighting the plurality of profound challenges and transformations permeating contemporary society and confronting the field of education. This leads to myriad pedagogical questions, all important and unique but also inseparable from each other in that they are manifestations of the complexity that is all around us. The challenges that confront us tend to be hard to define, open-ended, nonlinearly dynamic, nested, and interconnected. As educational researchers and teachers, therefore, we need not only ask ourselves how to respond pedagogically to the ecological crisis, the changing labor market, diversity issues, and so forth, but also the more fundamental question of how to educate our students in the face of the complexity that is all around us. What, in other words, does a pedagogical response look like that does justice to the experience and challenge of living in a complex world? My general concern is that – as an educational community – we fail to take seriously the complex reality we are dealing with and, consequently, come up with all too simplistic (e.g. reductionistic, superficial) solutions that reproduce the very predicaments we try to overcome. As the argument of the book unfolds, this concern shall be made more explicit. My hope is to contribute to a growing pedagogical movement aiming to foster a generative embrace of complexity and inspire teachers to join this cause.

If I am to take on this task, I need, first of all, to specify what "a pedagogical response" consists of and what kind of knowledge about such a response I aim for. The French pedagogue Philippe Meirieu (2016) developed a model that proves useful at this point. According to Meirieu, a pedagogical response ought to be developed along three interrelated dimensions. The first dimension is that of *axiology*: the purposes that provide a pedagogy its focus and orientation. The second dimension is that of *praxeology*: the educational settings, methods, teaching styles, and other practical arrangements that are utilized to shape the educational process. The third dimension is that of *theory*: the way both axiology and praxeology are grounded in premises, arguments, data, knowledge about the world.

S/he who is interested in pedagogy, in this view, embodies in him/herself the bridge between the educational philosopher, who is preoccupied with ideas about education, and the educational practitioner, who is preoccupied with the act of teaching. The consequence, as Meirieu emphasizes, is that the pedagogue is neither completely at home among the philosophers nor the practitioners, yet needs them both. I did, indeed, experience this tension in the process of working on this book. Numerous times have I found myself somewhat lost within the discourse of (educational) philosophers, who know more than I do about the landscape of philosophical theories and excel in speaking in reference of the great philosophers that preceded them. At such moments, I tend to feel the need to (re)connect the conversation to real experiences and challenges in a here-and-now sense. Numerous times, also, have I felt a certain restlessness when being among educational practitioners who so easily make practical choices, design curricula, and utilize methodologies. When this restlessness presents itself, I feel the need to pause to explicate, justify, and if necessary reformulate intentions and assumptions. When I act on such needs, my contribution is not necessarily appreciated by the philosopher or practitioner, or at least not always immediately. For, it tends to be somewhat of an interruption, slowing the practitioner down and challenging the philosopher to attend to here-and-now practical concerns.

To transform the awareness of this somewhat uncomfortable position of the pedagogue into an understanding of the task of doing pedagogical research, I find the work of Dutch pedagogue Gert Biesta helpful. Biesta (2016, 2018), argues that pedagogical research ought to be understood first and foremost in the context of its practical role. Its primary aim is, so he argues, not to explain or predict, nor to understand, but to support and improve the teacher's capacity to teach. Its orientation is, then, toward *helpful perspectives*, and it is exactly the researcher's position as both philosophical and practical that constitutes his/her potential helpfulness. Similar to the conceptualization of Meirieu, Biesta distinguishes two ways in which research can create knowledge that serves educational practice: (1) a technical way, which resonates with the domain of praxeology, concerning perspectives that can help teachers shape how to do things, and (2) a cultural way, resonating with the domain of axiology, concerning perspectives that can help teachers interpret education as an intentional cultural phenomenon. Building on these insights, my aim in this book is as follows:

To articulate and legitimize perspectives within the two interrelated domains of axiology and praxeology, that can help teachers to shape an adequate (i.e. complexity embracing) pedagogical response to the complex questions and challenges in our shared world.

What, then, does it mean for pedagogical perspectives to be helpful for teachers? It is, at this point, crucial to remember the open, emergent, and situated character of complex systems as described in this chapter, and consider if, in fact, we can understand educational systems and processes in these terms as well. So is, indeed, proposed by Biesta in his criticism of the dominance of quasi-causal thinking in contemporary educational research (Biesta 2010, 2016; see also Akkerman et al. 2021; Kincheloe and Tobin 2015). As Biesta argues, the educational process cannot be fully controlled by developing and utilizing effective techniques, for it is inherently open (i.e. there is no strict school-world dichotomy, this point is elaborated on in detail in Chap. 2), semiotic (i.e. based on meaning-making and interpretation), and recursive (i.e. the actions of teachers and students feed back into the educational process and influence its direction). There are, in this line of thought, no ready-made instructions that guarantee a desired outcome, and what desired outcomes are is highly debatable and culturally situated. My goal in this book, therefore, is to formulate helpful perspectives in a way that acknowledges the complexity of educational dynamics and helps teachers to make their own situated judgments. To do so, I understand helpful perspectives as hermeneutic lenses (Smedslund 2009); glasses teachers can put on and off in the ongoing effort to make sense of and improve their situated practices. Consider, for example, the two combined perspectives 'every student has personal interests that are worth pursuing' and 'personal interests can be triggered and deepened by teachers who notice and stimulate them'. These exemplary perspectives do not tell a specific teacher what the particular interests of a certain student are and exactly what would make that student be seen and stimulated. Yet, they do put exactly those questions at the center of a teacher's attention and can trigger a teacher to see, for instance, that a particular child is passionate about birds and would love to help distribute bird food around the school building in the winter (for further reading on this particular knowledge interest, I recommend Lengkeek 2016, p. 215–228; Smedslund 2009).

Modes of Inquiry

I now move on to an introduction of the three modes of inquiry this book builds upon. The first two modes are a direct consequence of positioning pedagogical research in the encounter of the educational philosopher and the educational practitioner and respectively engage with theoretical literature and real-life experiences of teachers. The third mode takes my own complicity (Davis and Sumara 2006) in educational innovation and research as its focus and is thus of a more personal nature. Let us look closer at these three modes of inquiry one by one.

Inquiry with Teachers as Co-researchers

As my aims in this book are of a creative and explorative nature – i.e. the generation of helpful pedagogical perspectives – an in-depth, longitudinal collaboration with a selected group of teachers holds more potential than, for instance, a large-scale survey study (which would be more suited for descriptive purposes). To involve teachers in this creative process, I have set out to find a group of teachers willing to participate, as *co-researchers* (Akkerman et al. 2021). Approaching teachers as co-researchers has strong implications. It means, first of all, that it matters who I ask and that it matters how I ask. Let me start with the former.

It is important that teachers as co-researchers share the overall concerns and questions of the research project they engage in. This asks, therefore, for a selective sampling strategy (Palinkas et al. 2015). In my case, teachers as co-researchers ought to recognize, in their own experiences and considerations, the aim to come up with an adequate pedagogical response to the complex questions and challenges that permeate the world-in-motion. Selecting co-researchers is, thus, a first step in focusing and organizing collaborative inquiry. The group needs to be purposefully homogeneous – in the sense of sharing a cause – and each teacher needs to be willing to commit to this inquiry in the capacity of co-researchers, meaning: willing to dive into one's own experiences, challenge one's ideas and assumptions, and engage actively in a collaborative, creative process of articulating helpful perspectives. However, within this homogeneous profile, it is desirable for there to be diversity in such variables as educational discipline and school type. Such heterogeneity is likely to enrich collaborative inquiry, for the tensions created by different backgrounds and experiences open up space for new insights.

To match this criterion of heterogeneity within purposeful homogeneity, in working on this book I have collaborated with a group of teachers I came to know through a project I previously participated in at The Bildung Academy (n.b. more detailed information about The Bildung Academy and my involvement in it follows when I introduce the third mode of inquiry). In the period September 2017 – May 2018 I co-initiated and hosted two pilot programs called Bildung for Teachers. The first of these pilots consisted of 7 weekly meetings and was attended by 5 teachers. The second pilot consisted of 8 two-weekly meetings and was attended by 9 teachers. These pilots intended to create an open space for teachers from diverse educational backgrounds interested in The Bildung Academy to explore each other's and the academy's intentions and approaches, and experiment in their classrooms with new, creative ideas. A central element of the intentions of The Bildung Academy is to create opportunities for students to personally engage with complex challenges in the world. This, indeed, strongly resonates with the aims of this book and it was for this reason that I expected that this group of teachers would share the concerns that drive my inquiry and would be willing to participate as co-researchers. At the same time, as the pilot program was aimed at teachers from any educational discipline, school type, and career phase, I knew this group of teachers to be of a naturally heterogeneous profile. To invite this group of teachers to join forces with me as co-researchers, I wrote a two-page invitation letter in which I described the aims and initial considerations driving the inquiry. In this letter, I also provided a rough sketch of the research process I intended to embark on together. Twelve of the 14 teachers responded positively to my invitation and thus became my co-researchers. The two teachers who did not participate were unavailable for health reasons. Table 1.1 provides a general profile of the diversity among these 12 teachers.

An important quality that working with teachers as co-researchers adds to this book, is the quality of exemplarity (Korsgaard 2019): it ensures that the helpful perspectives articulated throughout it are informed by and illustrated with exemplary teaching experiences that are relatable. It is important to state that at the moment that I ask a teacher to tell me about teaching experiences, I ask him or her to tell a story and use, therefore, a *narrative research approach* (Connelly and Clandinin 1990; Squire et al. 2013). As Connelly and Clandinin (1990) describe, such an approach recognizes people as 'storytelling organisms who, individually and socially, lead storied lives'. The study of narratives, thus understood, is 'the study of the ways

humans experience the world' (p. 2). My aim throughout this book is, therefore, not to generate factual descriptions of work conducted by my co-researchers, but rather to collaboratively inquire into the way they experience their work. As Polkinghorne (2007) summarizes (p. 479): 'storied texts serve as evidence for personal meaning, not for the factual occurrence of the events reported in the stories'. For this reason, communicative validation plays an important role throughout my research process, referring to occasions in which I feed narrative output back to a co-researcher to check if it matches the meaning s/he intended to narrate. It is, also, important to highlight here that the narratives that co-researchers articulate are to be understood as we-realities (Rosenthal and Fisher-Rosenthal 2004), that is: it is in interaction with me and other co-researchers that co-researchers' narrations take form. My complicity at this point consists of focusing co-researchers' attention on the considerations and concerns that drive the inquiry.

The focus on narrativity is congruent with the aim of articulating helpful perspectives, for having understood such perspectives as hermeneutic lenses they are narrative tools themselves. It is, however, important to distinguish between two types of narratives at this point. The first type is of a biographical nature (Kelchtermans 1993) and concerns teachers' narration of experiences. A second type is of the kind that I am ultimately interested in, namely hermeneutic lenses. This second type is of a more generic character and whereas biographical narratives are essentially historical, hermeneutic lenses can be used to look at the past, the present, and the future. Therefore, my work with teachers as co-researchers as I shall present it throughout this book combines, and alternates between, two steps of narration (Korthagen and Kessels 1999; Schön 1987). The first step is that of identifying and narrating personal teaching experiences. The second step is that of collaboratively engaging with these narratives in the attempt to interpret and articulate generic hermeneutic lenses.

Table 1.1 The general diversity profile of the 12 co-researchers at the start of participation

Gender			Age				School type		
Male	Female		25–40	41–50	51–65		High School	Applied University	Primary School
5	7		2	4	6		7	4	1
Location school (province)[a]			**Subject type**[b]						
North Holland	Utrecht	Groningen	Social Sciences	Humanities		Science		Art	Other[c]
5	3	3	3	4		2		3	1

[a]One participant is not listed here and came from the province of Gelderland
[b]The sum of subject types is 13 instead of 12 as one teacher taught in both Art and Social Sciences at a high school. Notably, during the research process this teacher switched jobs and started working as a teacher educator at an Applied University
[c]Other subject type refers to the primary school teacher

Literature Study

Just as the creative, explorative nature of my inquiry demands a purposeful selection of teachers as co-researchers, it suits my purposes to commit myself to particular strands of literature. To say it bluntly, if I wish to contribute to a pedagogical narrative in response to what I have called the confrontation with complexity, I need to rely on literature that acknowledges this complexity and takes it as a point of departure. I have, therefore, focused on two main types of scholarly work in this book. The first type is that of pedagogical theories that help to think about pedagogy from a complexivistic point of view. It can do so either in an explicit or implicit sense. Explicit complexivistic pedagogical theory refers to such scholars who in their pedagogical reflections actively frame education as a complex process in a transforming world. An example of such work is that of Crowell and Reid-Marr (2013), who developed a pedagogical account based on the concept of emergence, and in their reflections build forth extensively on such notions as interconnectedness and nonlinearity. Implicit complexivistic pedagogical theory, on the other hand, refers to such scholars who do not present themselves as complexity thinkers but in their pedagogical reflections, or at least in part of them, do develop pedagogical considerations that can be recognized as such. For instance, the educational reflections of Hannah Arendt (1961) can be considered complexivistic in the sense that for her education is ultimately attuned to the dynamic phenomenon of natality (i.e. through birth and decay the world is, and needs to be, constantly set right anew), yet the way she develops her arguments tends to fail to embrace the complex transition from childhood to adulthood and, particularly, how young people already can be seen as co-shapers of the world they live in. Or, to provide another example, the work of Paulo Freire (1972) can be seen as complexivistic in the sense that he considers the process of liberation as one which demands dialogical action and reflection involving both the oppressed and the oppressor, but simultaneously his work is informed by a humanistic worldview which obscures from view the deep interdependencies between human and nonhuman forms of life.

The second type of scholarly work is such theoretical reflections that provide a deeper understanding of the ontological implications of a complexivistic worldview. The work of Davis and Sumara (2006) and Wahl (2016) are examples of this type of literature that already crossed my path in Chap. 1. In the chapters to come, my engagement with this type of literature will be deepened and expanded. For, in order to contribute to types of pedagogy that engage meaningfully with the complex questions and challenges in the world-in-motion, it is crucial to build an understanding of what (young) human beings' relationship to such challenges and questions is in the first place.

There is, of course, a vast amount of literature within these categories that could potentially enrich this book. How, then, am I to go about selecting literature? Two considerations have informed my strategy at this point. Firstly, it is important to remember the purpose of literature study in this book as serving the articulation of helpful pedagogical perspectives. My engagement with literature is relevant in so far as it contributes to articulating perspectives, but is not an end in itself. It would be somewhat overdoing it to expect myself to read all the potentially relevant literature out there, and to give an extensive overview hereof, just as it would be overdoing it to include all teachers with relevant experiences to my inquiry with co-researchers. Just as with co-researchers, I consider it more important that my inquiry is qualitatively rich – i.e. that I engage with literature which is excellent in its field and which truly deepens understanding of matters crucial to this book – than that it is quantitatively exhaustive in the sense of including everything that can potentially contribute. Secondly, I have had to deal, unavoidably, with practical limitations. Especially in combination with the other two modes of inquiry, the time I have had at hand for literature study forced me to be selective. This, also, is part of the somewhat uncomfortable position of the pedagogical researcher that I spoke of in reference to Meirieu (2016): I want to engage with literature and with teachers simultaneously, but due to practical constraints am limited in the time and energy I can devote to each.

Following these two considerations, I have opted for a pragmatic approach to literature selection. There have been two main ways in which I selected literature for this book. The first way resembles what is commonly known as snowball sampling (Noy 2008). This refers to occasions when initial writings or presentations resonated in such a way with other researchers that they suggested specific literature that they considered of added value for my cause (e.g. during conference visits, meetings with supervisors, or informal dialogue with colleagues and acquainted researchers), or occasions in which literature that I was reading lead me to other literature that proved particularly interesting for my purposes. The second way is data-driven and refers to occasions when inquiry with co-researchers illuminated certain insights or perspectives that focused attention on particular texts. Sometimes this happened in such a direct way that either a co-researcher or I would remember and suggest particular publications that we believed might help to deepen specific insights. On other occasions, this would happen in a more indirect way, when I felt a need to search for new literature that zooms in on a specific concept or perspective that came up during one or multiple sessions.

Autoethnography

The initiative for this book is rooted, for a great deal, in my own recent experiences in the field of education. In February 2015, I became one of the initiators of a student-driven educational initiative in the Netherlands: The Bildung Academy (see Initiators of The Bildung Academy 2017). Like me, most of my fellow initiators were still students at the time (I graduated with a research master's in educational sciences later that year) and the source of the initiative lay in a shared frustration built during our student years. Elsewhere, I summarized this frustration as follows (Wessels 2017, p. 16, freely translated):

As students, we feel short-changed, for we seldomly have to show who we are. We are surprised that we don't learn actively a lot, that we don't meet many people with other viewpoints, that we don't have to provide our viewpoints often, that we don't have to leave our comfort zones regularly, that we rarely need to reflect on the systems we are part of.

This shared frustration was particularly situated in an awareness that we live in 'a complex world in transition' (Initiators of The Bildung Academy 2017, p. 5). We felt, indeed, that higher education could do a great deal more to support students 'to acquire a nuanced understanding of complex issues' (p. 2) and 'to actively take a position in this world' (p. 5). Based on these experiences, we founded The Bildung Academy and started offering courses focusing on what we experienced as pressing themes in our contemporary society, such as the changing role of religion and spirituality, the role of money in our lives, the transition toward renewable energy, the emergence of circular economy, and the digitalization of more and more aspects of our lives. In these courses, we aimed to create a context for students to explore these themes together and in interaction with a wide range of relevant others (e.g. researchers, entrepreneurs, artists, activists, policymakers), and to work toward some sort of own action or initiative aimed at making a difference that matters.

To our amazement, the initiative quickly took off as our programs filled with interested students, we acquired supportive funding and took attention from various schools and universities in the Netherlands. Today, The Bildung Academy is an established organization hosting numerous experimental educational programs and collaborating with educational institutions throughout the Netherlands, offering strategical advice, training teachers, and helping (re)design curricula. For me personally, this whole process has been a truly transformative experience, for it taught me the invaluable lesson that through my actions I can make a real difference in the systems I am part of and thus take responsibility for their ongoing (re)generation. I learned, in other words, that the educational system is not stationary but in motion and that I am a movement in it.

It is this experience, in combination with the frustrations and intentions behind the initiation of The Bildung Academy as just described, that moved me to start working on this book. Through pedagogical inquiry, I hope to deepen my commitment to the educational field. The helpful perspectives developed throughout this book are intended to help educational researchers and teachers, but I too hope to utilize them in experimental educational endeavors to come.

Since the initiation of The Bildung Academy, I have developed and taught in several experimental educational programs for students and teachers, I have given numerous talks and workshops on various occasions to educational professionals, and as an educational consultant, I have been involved in several projects aimed at curricular innovation and teacher training in the Netherlands whilst working on this book. The teachers I work with as co-researchers are, too, part of the Bildung Academy network, and to me, collaborative inquiry with them is part of my engagement with educational practice. To utilize such personal experiences in this book, I engage in autoethnography (Ali-Khan 2015; Ellis et al. 2011). Doing autoethnography entails to 'retrospectively and selectively write about epiphanies that stem from, or are made possible by, being part of a culture and/or by possessing a particular cultural identity' (Ellis et al. 2011, p. 276). Autoethnography 'takes seriously the idea that the study of self is legitimate' (Ali-Khan and White 2019, p. 741); yet, this legitimacy is grounded in the researcher's ability to transparently and lively narrate his/her experiences and findings and his/her effort to bring these into dialogue with other practitioners and existing theoretical frameworks (Ellis et al. 2011). In this book, therefore, I use autoethnography to contribute to the qualities of exemplarity and transparency by bringing it into dialogue with my other two modes of inquiry.

In the chapters to come I will provide autoethnographic accounts on several occasions, zooming in on triggering experiences in my work for TBA and in facilitating the inquiry with co-researchers. As is common in autoethnography (Ellis et al. 2011), I focus herein on those experiences and insights that for me are most striking

and transformative. To facilitate this, I have utilized the following strategies: (1) the collection of personal notes of triggering events, insights, thoughts, and emotions experienced throughout the research process, (2) writing yearly process reports depicting most triggering experiences and gained insights, and discussing these with colleagues, and (3) gathering feedback of co-researchers regarding how they experience the research process and my role in it.

Moving Forward

Now that I have introduced the background and aims of this book, as well as the three threads of inquiry it builds on, it is time to embark on a journey. In what follows, I shall present my journey step by step and in doing so I shall further develop and legitimize my methodological approaches. I hope that the introduction of the three modes of inquiry has made it clear that they are not separate but interwoven and not hierarchically structured but on equal terms with each other, for that is how I intend to treat them. Every mode of inquiry feeds back to the other two modes. In the chapters to come, theoretical concepts derived from literature study help frame and focus inquiry with co-researchers and autoethnography, which in their turn trigger the study of particular literature. Likewise, autoethnography helps explain my interests and focus to co-researchers, and the process with co-researchers, in turn, triggers autoethnographic reflection on my behalf. The relationship between the modes of inquiry is, thus, of a bi-directional nature, and it is in their mindful interweaving that I work on the generation and articulation of helpful pedagogical perspectives. I hope that you will find it a stimulating read!

References

Akkerman, S. F., Bakker, A., & Penuel, W. R. (2021). Relevance of educational research: An ontological conceptualization. *Educational Researcher,* 50(6), 416–424.

Alhadeff-Jones, M. (2008). Three generations of complexity theories: Nuances and ambiguities. In Mason,

M. (Ed.), *Complexity theory and the philosophy of education* (p. 62–78). Hoboken, NJ: Wiley Blackwell.

Ali-Khan, C. (2015). Liberation, mice elves and navel gazing: Examining the ins and outs of autoethnography. In Tobin, K., & Steinberg, S. R. (Ed.), *Doing educational research: A handbook* (2nd ed.) (p. 293–319). Rotterdam: Sense Publishers.

Ali-Khan, C., & White, J. W. (2019). Between hope and despair: Teacher education in the age of Trump. *Educational Philosophy and Theory,* 52(7), 738–746.

Arendt, H. (1961). The crisis in education. In H. Arendt (Ed.), *Between past and future: Six exercises in political thought* (p. 173–196). London: Faber and Faber.

Banks, J. A. (2008). Diversity, group identity, and citizenship education in a global age. *Educational Researcher,* 37(3), 129–139.

Beckert, J. (2016). *Imagined futures: Fictional expectations and capitalist dynamics.* Cambridge, MA: Harvard University Press

Besley, T., & Peters, M. A. (2019). Life and death in the Anthropocene: Educating for survival amid climate and ecosystem changes and potential civilisation collapse. *Educational Philosophy and Theory,* 52(13), 1347–1357.

Biesta, G. J. J. (2010). *Good education in an age of measurement: Ethics, politics, democracy.* Boulder, CO: Paradigm Publishers.

Biesta, G. (2016). Improving education through research? From effectiveness, causality and technology to purpose, complexity and culture. *Policy Futures in Education,* 14(2), 194–210.

Biesta, G. (2018). *Tijd voor pedagogiek: Over de pedagogische paragraaf in onderwijs, opleiding en vorming.* Utrecht: Universiteit voor Humanistiek.

Bond, M., Marin, V. I., Dolch, C., Bedenlier, S., & Zawacki-Richter, O. (2018). Digital transformation in German higher education: Student and teacher perceptions and usage of digital media. *International Journal of Educational Technology in Higher Education,* 15(1), 1–20.

Connelly, F.M., & Clandinin, D.J. (1990). Stories of experience and narrative inquiry. *Educational Researcher,* 19(4), 2–14.

Coyne, R. D. (2005). Wicked problems revisited. *Design Studies,* 26(1), 5–17.

Crowell, S., & Reid-Marr, D. (2013). *Emergent teaching: A path of creativity, significance, and transformation.* Lanham, MD: Rowman & Littlefield.

Davis, B., & Sumara, D. J. (2006). *Complexity and education: Inquiries into learning, teaching, and research.* New York, NY: Routledge.

Dixson, A. D. (2017). "What's going on?": A critical race theory perspective on Black Lives Matter and activism in education. *Urban Education,* 53(2), 231–247.

Ellis, C., Adams, T. E., & Bochner, A. P. (2011). Autoethnography: an overview. *Historical Social Research,* 36(4), 273–290.

Freire, P. (1972). *Pedagogy of the oppressed.* Harmondsworth: Penguin.

Hajir, B., & Kester, K. (2020). Toward a decolonial praxis in critical peace education: Postcolonial insights and pedagogic possibilities. *Studies in Philosophy and Education,* 39(5), 515–532.

Harari, Y.N. (2016). *Homo Deus: A brief history of tomorrow.* London: Harvill Secker.

Heikkilä, M., & Mankki, V. (2021). Teachers' agency during the Covid-19 lockdown: A new materialist perspective. *Pedagogy, Culture & Society.* Advance online publication. https://doi.org/10.1080/14681366 .2021.1984285.

Hoffman, J., Pelzer, P., Albert, L., Béneker, T., Hajer, M., & Mangnus A. (2021). A futuring approach to teaching wicked problems. *Journal of Geography in Higher Education.* Advance online publication. https://doi.org /10.1080/03098265.2020.1869923.

Initiators of the Bildung Academy (2017). *De Bildung Academie manifesto English version.* Retrieved 11-10-2021 from https://uploads-ssl.webflow.com/60b0bb b1dfea697125ee3878/60ddb6f8e6db6cbe3fe4b154_ Manifesto-ENG.pdf.

Ingold, T. (2008). Bindings against boundaries: Entanglements of life in an open world. *Environment and Planning A: Economy and Space,* 40(8), 1–15.

Kelchtermans, G. (1993). Getting the story, understanding the lives: From career stories to teachers' professional development. *Teaching and Teacher Education,* 9(5/6), 443–456.

Kincheloe, J. L., & Tobin, K. (2015). Doing educational research in a complex world. In Tobin, K., & Steinberg, S. R. (Ed.), *Doing educational research: A handbook* (2nd ed.) (p. 3–13). Rotterdam: Sense Publishers.

Korsgaard, M. T. (2019). Exploring the role of exemplarity in education: Two dimensions of the teacher's task. *Ethics and Education,* 14(3), 271–284.

Korthagen, F. A. J., & Kessels, J. P. A. M. (1999). Linking theory and practice: Changing the pedagogy of teacher education. *Educational Researcher,* 28(4), 4–17.

Laloux, F. (2014). *Reinventing organizations: A guide to creating organizations inspired by the next stage of human consciousness.* Brussels: Nelson Parker.

Laszlo, E. (1996). *The systems view of the world: A holistic vision for our time.* Cresskill, NJ: Hampton Press.

Lengkeek, G. (2016). *Pedagogisch leiderschap: het ondersteunen van vorming door onderwijs in exacte vakken.* Delft: Uitgeverij Eburon.

Lotz-Sisitka, H., Wals, A. E. J., Kronlid, D., & McGarry, D. (2015). Transformative, transgressive social learning: Rethinking higher education pedagogy in times of systemic global dysfunction. *Current Opinion in Environmental Sustainability,* 16, 73–80.

Meirieu, P. (2016). *Pedagogiek: De plicht om weerstand te bieden* (S. Verwer, Vert.). Culemborg: Uitgeverij Phronese.

Morrison, K. (2008). Educational philosophy and the challenge of complexity theory. In Mason, M. (Ed.), *Complexity theory and the philosophy of education* (p. 16–31). Hoboken, NJ: Wiley Blackwell.

Noy, C. (2008). Sampling knowledge: the hermeneutics of snowball sampling in qualitative research. *International Journal of Social Research Methodology,* 11(4), 327–344.

Palinkas, L. A., Horwitz, S. M., Green, C. A., Wisdom, J. P., Duan, N., & Hoagwood, K. (2015). Purposeful sampling for qualitative data collection and analysis in mixed method implementation research. *Administration and Policy in Mental Health and Mental Health Services Research*, 42(5), 533–544.

Peters, M.A. (2017). Education in a post-truth world. *Educational Philosophy and Theory*, 49(6), 563–566.

Polkinghorne, D. E. (2007). Validity issues in narrative research. *Qualitative Inquiry*, 13(4), 471–486.

Rittel, H. W. J., & Webber, M. M. (1973). Dilemmas in a general theory of planning. *Policy Sciences*, 4, 155–169.

Rosenthal, G., & Fisher-Rosenthal, W. (2004). The analysis of narrative-biographical interviews. In Flick, U., Kardoff, E., & Steinke, I. (Ed.), *A companion to qualitative research* (p. 275–281). London: SAGE Publications.

Schön, D. A. (1987). *Educating the reflective practitioner: Toward a new design for teaching and learning in the professions*. San Fransisco, CA: Jossey-Bass.

Scott, L. A. (2017). *21st Century skills early learning framework*. Partnership for 21st Century Learning, retrieved 24-09-2018 from http://www.p21.org/our-work/elf.

Smedslund, J. (2009). The mismatch between current research methods and the nature of psychological phenomena: What researchers must learn from practitioners. *Theory & Psychology*, 19(6), 778–794.

Son, C., Hegde, S., Smith, A., Wang, X., & Sasangohar, F. (2020). Effects of COVID-19 on college students' mental health in the United States: Interview survey study. *Journal of Medical Internet Research*, 22, 1–14.

Squire, C., Andrews, M., & Tamboukou, M. (2013). Introduction: What is narrative research? In Squire, C., Andrews, M., & Tamboukou, M. (Ed.). *Doing narrative research* (2nd ed.) (p. 1–26). London: SAGE Publications.

Tatham-Fashanu, C. (2021). A third space pedagogy: Embracing complexity in a super-diverse, early childhood education setting. *Pedagogy, Culture & Society*. Advance online publication. https://doi.org/10.1080/14681366.2021.1952295.

United Nations (2015). *Goal 4: Ensure inclusive and equitable quality education and promote lifelong learning opportunities for all*. Retrieved 01-02-2022 from https://sdgs.un.org/goals/goal4

Wahl, D. C. (2016). *Designing regenerative cultures*. Axminster: Triarchy Press.

Walburg, V. (2014). Burnout among high school students: A literature review. *Children and Youth Services Review*, 42, 28–33.

Wessels, K. R. (2017). *Dan maken we ons onderwijs zelf wel: Een bildungsvisie*. Leusden: ISVW uitgevers.

Zembylas, M. (2020). The affective grounding of post-truth: Pedagogical risks and transformative possibilities in countering post-truth claims. *Pedagogy, Culture & Society*, 28(1), 77–92.

With the help of Meirieu (2016), I have specified my aim in this book to contribute to a pedagogical response to complex questions and challenges in contemporary society along the dimensions of axiology (i.e. the purposes that provide a pedagogy its focus and orientation), praxeology (i.e. how the educational process is shaped), and theory (i.e. premises, arguments, data, knowledge about the world). Before I focus my attention on the dimensions of axiology and praxeology, I consider it important to expand my understanding of complexity thinking as introduced in Chap. 1 and thus zoom in on the dimension of theory. Particularly, my considerations thus far have focused on ways in which societal challenges can be understood to be complex, but I have not explicitly devoted attention yet to the question of students' relationship to them. As a teacher, in other words, how am I to understand the relationship that my students have to such challenges as the ecological crisis, post-truth politics, cultural diversification, and so forth? Such an understanding is crucial for my purposes, for it is this very relationship that the pedagogical response I seek to contribute to is to be directed at.

In this chapter, therefore, I present an initial inquiry – along the three methodological threads of literature study, autoethnography, and inquiry with co-researchers – aimed at identifying a complexivistic ontological premise (i.e. an understanding of the nature of students' relationship to complex societal challenges) that can open up a subsequent inquiry into the pedagogical dimensions of axiology and praxeology. First, utilizing literature study and autoethnography, I develop the initial understanding that students are not only to be approached as in-preparation-for-future-participation but also as participants in the ongoing, collaborative effort to learn about complex societal challenges and work toward meaningful solutions. Second, I use this initial understanding to start up the process of inquiry with my co-researchers and collect exemplary teaching experiences in which students actively participate in complex societal challenges. Third, I share a pattern of insight that emerged through this process – the insight that students are always already uniquely part of complex societal challenges through their evolving biographies – and build on, especially, the relational ontologies of Tim Ingold and Karen Barad to propose to understand students' relationship to complex societal challenges in terms of entanglement.

The Readiness Dichotomy and the Student as Participant

In Chap. 1 I emphasized – in agreement with Rittel and Webber (1973), Davis and Sumara (2006), and Wahl (2016) – that as complex challenges are hard to define, multi-interpretable, open-ended, and dynamic, they cannot be solved or mastered a priori and once and for all, but rather demand an ongoing engagement, attentive to what emerges in the here and now. In what follows, I argue that this logic sheds an interesting light on the common sense understanding of education as a process of preparation or qualification (Biesta 2010, 2020). This leads, in turn, to an initial understanding of how we might conceptualize the relationship of students to complex societal challenges. Allow me to elaborate.

It is rather intuitive to understand education as a process of preparation. After all, education tends to give us access to things previously out of reach. If we complete primary school, we are considered ready for secondary school, the completion of which provides access to vocational training or university, which in turn enables us to take on particular positions in the labor market. This system of stages-of-preparation plays a crucial role in guaranteeing quality in many professional practices, ensuring that doctors save lives every day, that psychologists can be trusted to help us face our deepest fears and struggles, that architects and construction workers create safe buildings to live in, and so forth. Likewise, the rationale of preparation plays a crucial role in democratic societies, as the right to vote is given to citizens only after having received citizenship education at school and having reached a certain age. Citizenship education, following this logic, is where we ensure that the newcomers in our democracies develop the knowledge, beliefs, and competencies that befit participation in it (n.b. this inevitably raises the question who decide what these are and how that decision is made). Yet the logic of preparation also exists in more progressive forms, emphasizing – rather than the necessity of conserving that which we value today – the importance of the renewal that new generations might bring about in the future. Exemplary for such an approach – in the context of democratic citizenship – is Hannah Arendt's educational critique (1961), in which she argues that education is ultimately attuned to the phenomenon of natality, 'the fact that we have all come into the world by being born and that this world is constantly renewed through birth' (p. 196). Arendt considers this renewal crucial, as 'to preserve the world against the mortality of its creators and inhabitants it must be constantly set right anew' (p. 192). Therefore, she argues that it is the teacher's task 'to preserve this newness' (i.e. the teacher should leave it to the student what kind of renewal s/he shall bring about) and 'to introduce it as a new thing into an old world' (p. 193). Underlying her argument – and in this sense, Arendt's reasoning does not fundamentally differ from more conservative interpretations of citizenship education – Arendt poses a separation between what she calls the educational realm and the political realm. In Arendt's educational realm, newcomers are introduced into the world while preserving their newness, and post-education they enter the political realm, ready to act in the world and take responsibility for it.

Although preparative approaches thus often serve the interests of the labor market and/or democracy, historically it was in fact the sake of the child itself, and a commitment to its safe and healthy development, that played a crucial role in the emergence of the preparative logic (Dasberg 1975). In her historical analysis, Dasberg – who was an influential Dutch pedagogue in the 1970s and 1980s – describes how the awareness grew during the Age of Enlightenment that children have specific needs of their own and, therefore, in education, in literature, on the labor market, and in law should be treated differently than adults. In other words, the field of developmental psychology – the origin of which is often traced back to 1762, when Jean-Jacques Rousseau published his famous *Emile: Or, on education* (1979) – and the deliberation of its far-reaching implications, started emerging. Consequently, Dasberg argued, "Youthland" got created: 'the area – an area in time – in which the child can ripen in a community of peers who are in the same developmental phase' (freely translated, p. 19). A preparative approach to education, thus, can be differentiated across many contexts (e.g. preparing for the labor market, democratic citizenship, adulthood) and ideologies (e.g. attuned to the conservation and/

or ongoing renewal of certain ideals or quality standards) yet what binds all such approaches together is that they acknowledge and welcome students – in one way or another – as not being ripe or ready yet and provide them with the time, context, and support to prepare for a next phase or step in life.

Interesting for our concerns in this chapter is that this logic of preparation thus installs dichotomies in terms of "readiness". Dichotomies, between those who can, know, or have certain rights (i.e. the teacher, and those particular others the teacher represents), and those who are still unfit, lack knowledge, and are yet to deserve certain rights (i.e. students). Education, consequently, is the process of resolving the lack of readiness, and the core concern of educational research becomes, thereby, which particular lack of readiness is to be resolved and how this can and should be achieved. If we strictly apply this preparative logic to the relationship of students to complex societal challenges, the conclusion is that they are in some phase of preparation for future participation in them. Of course, there is a truth to this in the obvious sense that, for instance, it can be expected to be too much of a challenge and responsibility for a student of Environmental Sciences to take on the role of Environmental Minister. Yet, if we look back at the complexivistic understanding with which I started this section, it is not hard to see that an understanding of students' relationship to complex societal challenges in purely preparative terms would be insufficient. For, in the context of complex societal challenges, it is impossible to define the point at which someone is done learning and thus fully ready. Even our ministers and most accomplished experts are not in such a luxurious position. No matter how prepared we are, when we are dealing with complex societal challenges we need to learn and respond in the moment and a strict readiness dichotomy, therefore, is insufficient.

We might consider two ways in which the insufficiency of a strict readiness dichotomy manifests itself in the face of complexity. First of all, in this context a distinction in readiness between a learner and an expert, or between a student and a teacher, is always relative rather than absolute; although teachers evidently embody valuable knowledge and experiences that students can benefit from, they too find themselves in need of knowledge and experience in the face of complexity. Following complexity thinking, teachers and experts are – and need to be – learners as well (Crowell and Reid-Marr 2013), or as Freire (1972) proposed, teachers are teacher-students. Secondly, readiness-dichotomies tend to be reversible in particular ways, depending, of course, on who is involved in which context. Perhaps the most familiar images hereof in contemporary society are those of a teenager teaching his/her (grand)parent how to use some piece of innovative technology (e.g. a new phone or computer program) and of global student climate strikes in which the young of our world inspire older generations in their commitment to transitioning to more sustainable ways of living. I agree with Freire, therefore, that not only should we understand teachers as teacher-students, but also would it do justice to students to consider them student-teachers, or as he formulates his conclusion (1972, p. 45): 'education must begin with the solution of the teacher-student contradiction, by reconciling the poles of the contradiction so that both are simultaneously teachers and students'.

The complexity we face demands of us, so is my initial argument, to nuance the readiness-dichotomy, to allow students and teachers alike to identify themselves with, and act from the position of, both sides of the dichotomy as they continue to learn together. To illustrate this point, I would like to end this reflection on the readiness dichotomy by sharing an experience from my work for The Bildung Academy. This experience involves a group of teachers and students that together de-construct the solidified dichotomy between them:

> April 2019, Amsterdam. I facilitated a session with teachers and students in public administration at a University of Applied Sciences, initiated by teachers who were dissatisfied with their students' engagement. To start, I invited students and teachers to share their feelings and desires, which were diverse and mirrored each other. Some teachers wanted students to take more initiative, whilst some students wanted teachers to take them on a

journey. Vice versa, some students wanted more personal attention and freedom to pursue their own interests, whilst some teachers desired their students to simply trust in and commit to the educational material they offer. When collecting insights afterward, both students and teachers emphasized that framing the issue at hand as "low student engagement" felt like an unfair localization of the real issue and decided to formulate a question that turns students and teachers into partners in a shared cause: how can we be more engaged together? For the rest of the session, students and teachers worked together with renewed energy and openness, and they came up with several initial plans to revitalize their joined educational experience. I feel like I have been part of a powerful, revitalizing experience of freeing fixed and opposed student- and teacher positions.

From the perspective of complexity thinking, thus, it is important to not only understand the relationship of students to complex societal challenges as in-preparation-for-future-participation but also in terms of participation-here-and-now, as partners in the ongoing, collaborative effort to learn about complex societal challenges and work toward meaningful solutions. Such is, then, my initial understanding of students' relationship to complex societal challenges, and in the next sections I will explore this participatory-perspective further. Yet before I go there, allow me to spend a few words on educational policy in our contemporary world. For, in light of these considerations, it somewhat worries me to notice numerous early twenty-first-century analyses pointing toward a focus in educational policy and research on well-defined skills and knowledge and their controlled measurement (especially Biesta 2010; Nussbaum 2010; Olssen and Peters 2005), as well as the not infrequent proposals in our field to add skills and competencies to already extensive lists of learning outcomes. Perhaps the most striking example hereof is the twenty-first-century skills movement (Scott 2017; see also the critique of Biesta 2013b), which proposes a set of 12 skills the development of which is to ensure that students can stay competitive in a changing labor market. If we are to take the complexity that confronts us seriously, we need to be careful with this kind of preparatory-reflex, for if driven too far it is an attempt to predict, control and

tame a complex reality which is doomed to be insufficient. Luckily, there are reasons to be optimistic as well. For backed by influential critiques such as those of Freire (1972), Biesta (2010), and Nussbaum (2010), the critique of a strictly preparative approach to education has become widespread. To provide two examples for illustrative purposes (indeed many more could be added): (1) the idea and practice of community service learning, which enables students to participate in, and reflect on, services that both benefit the public and contribute to their own development, is receiving increasing attention (Tijsma et al. 2020), and (2) Reid et al. (2008) analyzed a trend in the fields of health- and sustainability-promoting education to move from an initially moralistic to a more participatory, democratic approach. Interestingly, also, Dasberg's historical analysis (1975) describes not only the emergence of preparative approaches to education but also how this triggered the subsequent emergence of the argument that by strictly separating children from the adult world we problematize their healthy development and throw away the rich learning opportunities that reside in a pedagogically mediated encounter with contemporary challenges. Today, Dasberg's suggestion that participation and first-hand experience is a developmental need, is convincingly backed up by accumulated insights from the field of constructive developmental psychology (Eriksen 2006; Kegan 2018) and emerging insights in the field of embodied cognition (Anderson 2003; Shapiro and Stolz 2019).

Educational Space for Participation – An Exploration

It was at the point in time of working on this book that I had developed the initial argument thus presented that I started up my collaboration with teachers as co-researchers, and my purposes in doing so were twofold. My first purpose was to explore exemplary teaching experiences in which students actively participate in contemporary complex challenges so that a better understanding of students' relationship to such challenges in participatory terms might emerge. My second purpose was to kick-start the process of inquiry with co-researchers by creating rich biographical narratives that can inspire subsequent efforts to collaboratively interpret and articulate helpful perspectives in the dimensions of axiology and praxeology (see Chap. 1). To match these purposes, I decided to start my collaboration with co-researchers by conducting a narrative-biographical semi-structured interview (after Kelchtermans 1993) with each of them. These interviews took place between February 2019 and February 2020, took 2.5–3 h each, and were structured as follows.

Step 1. Co-researcher fills out a form asking for a schematic overview of past teaching experiences (i.e. when, where, what type of school, which student population, which subjects) and other work experiences that – in his/her opinion – are closely connected to his/her work as a teacher.

Step 2. Conversation 1: pedagogical purpose; what is your pedagogical intention as a teacher?; what do you hope to accomplish through education?

Step 3. Introduction of the focus of this session on exemplary experiences in which students participate in questions and challenges that permeate contemporary society.

Step 4. Conversation 2: exemplary experiences; following the introduction on the focus of this session, which of your own teaching experiences come to your mind first?

Step 5. Conversation 3: zooming in on a particularly inspirational exemplary experience of the co-researcher's choice; what happened? what was your role? how would you describe your students in this experience?

Step 6: Conversation 4: zooming in on a particularly frustrating exemplary experience of the co-researcher's choice; what happened? what was your role? how would you describe your students in this experience?

Step 7: Conversation 5: closing reflections; which insights did this session trigger?

I further structured each of the five conversational rounds with the following sub-steps. First, I would introduce the focus of a conversational round and introduce the relevant guiding questions to start the conversation. Then, a conversation would evolve in which I would take on the role of listening attentively and asking open follow-up questions to invite more precise and/or elaborate answers. After the conversation, I would then ask the co-researcher to take a moment for him/herself to briefly write down on a complementary form what s/he perceived to be the core content of that conversational round. Together with audio recordings of each conversational round, these summary notes were important input for the next stage of the research process. After each interview, namely, I would process the data generated through this procedure into what Kelchtermans (1993) calls a narrative synthesis text; a lively narration of a co-researcher's experiences and reflections following the structure of the interview and adhering to the co-researcher's choice of words. Typically, these narratives would be 3 or 4 pages in length. To ensure communicative validity, I would ask co-researchers to read their texts and to suggest alterations if needed, typically leading to a few minor alternations or additions.

The 12 biographical narratives that this procedure resulted in, are input for the next stages of inquiry with co-researchers, as will be presented in Chap. 3. In the current chapter, however, my interest is in the kind of exemplary experiences narrated by my co-researchers and if these, perhaps, point toward a deeper understanding of students' relationship to complex societal challenges. For that purpose, in Table 2.1 I present a brief overview of two exemplary experiences per co-researcher, selected in such a way as to limit overlap and show a rich picture (n.b. the names used in Table 2.1 are not co-researchers' real names).

Table 2.1 Exemplary experiences of space for participation

Teacher	School type and subject	Exemplary experiences
Ronald	HS[a]; social studies, theatre	Students sharing personal experiences of informal caregiving followed by a group discussion on the desirability hereof
		Co-creating a musical with students about sexting and performing it for the whole school, parents, and other invitees.
Jacob	HS; physics, informatics	Weekly classroom discussions with students about infographics revolving around societal issues
		Supervising and helping students formulate their own research questions and graduation projects
Elmarie	HS; Dutch language and culture	Students designing each other's reading literacy exams with self-chosen news articles, leading to a lively classroom discussion about a certain news article about the ethicality of an Anne Frank themed escape-room
		Adapting the curriculum to student climate change protests by using articles about it in class and facilitating group discussions about the theme
Astrid	HS; Dutch language and culture	Third graders designing a buddy system for first graders in response to experienced challenges in transitioning from primary school to high school
		Students uttering extreme (e.g. racist) statements in the classroom, leading to a tense vibe and classroom discussion about the sources and impact of such statements
Sandra	UAS[a]; marketing	Building marketing courses around students' personal experiences with brands
		Co-creating teaching method with students
Aafke	UAS; fashion	Open dialogue with students about frustrations with the educational system, mismatching expectations, and how to deal with such experiences
		Students collecting and analyzing inspirational material (e.g. film, text, pictures, etc.) to recognize core themes that particularly engage or touch them
Anika	HS; English and Spanish language and culture	Welcoming an exchange student from Costa Rica into the classroom with personal video's about life in the Netherlands
		Reading 'The absolutely true diary of a part-time Indian' together and relating core themes in it, such as cultural diversity, bullying, and poverty, to students' own experiences through individual assignments and group discussions
Irene	HS; worldview education	Students creating education for each other about self-chosen existential themes such as loneliness and suicidality
		Dialogues with students from diverse religious backgrounds about religious beliefs during a school trip to Vatican City.
Jens	UAS; Law & ethics of law	Discussing recent law cases with students and relating them to personal experiences, leading to discussions about the limits of responsibility and accountability.
		Co-teaching a course for students from other faculties designed around the idea of open-ended critical reflection on the law system, leading to self-initiated student projects
Kasper	UAS; microbiology	Supporting a student in supervisory meetings who struggles to deal with an intercultural communication conflict with a lab colleague during an internship
		Discussing a film about research ethics with students

(continued)

Table 2.1 (continued)

Teacher	School type and subject	Exemplary experiences
Pien	PS[a]	Students wanting to initiate charity for whales
		Trying to make a boy who recently fled to the Netherlands from Eritrea, a boy diagnosed with ADHD, and a boy struggling with anger management work together and articulate their thoughts through philosophical assignments for children
Steven	HS; Drawing	A student interrupts a lecture, stating "seen from my cultural background you misinterpret orientalism" and takes the stage to explain her perspective
		Steven finds out that one of his students is a nephew of one of his favorite artists, and decides to improvise a class about this artist's life and work in collaboration with the student

[a]UAS = University of Applied Sciences; HS = High School; PS = Primary School

The Space Dichotomy and the Entangled Student

As the research process with my co-researchers progressed, an important pattern of insight for my current concerns started emerging. In our conversations focused on exemplary experiences, the following perspective surfaced repeatedly: in their own ways, students always already are part of complex societal challenges through their evolving biographies. To put it in the words of two of my co-researchers (respectively Aafke and Ronald): 'students who enter the school always already are someone in the world, every student has a background in certain family-, cultural-, and educational systems and moves within certain systems today', and 'within every group of students there is such a richness of characters and backgrounds, [...] students need to connect themes to their own identity and background and have their own themes and concerns that can serve as a point of departure in education.' To illustrate this point, I would like to provide a more detailed description of three exemplary experiences of co-researchers depicted in Table 2.1:

> In preparing them for upcoming graduation exams, high school drawing teacher Stephen gives his students an art history lecture about orientalism. Whilst doing so, a student raises her hand and shares her opinion that they are approaching orientalism through the modern lenses of the West. She

goes on to confront Stephen and her classmates with the Western representation of the unknown, the shrouded woman, and the harem, and how these can be approached differently. Stephen decides to give her space, thinking to himself "you are going to teach us something which I can't". Stephen then takes on the role of conversation moderator and later on joins in the conversation by trying to explicate the art-historical perspective that he is an expert on.

Worldview education teacher Irene provides her 5-vwo students (i.e. the fifth out of 6 years of pre-university education) with the opportunity to host sessions for each other on self-chosen existential themes. The two students whose turn it is, somewhat astonish her as they put forward the themes of loneliness and suicidality in a manner more direct than she would dare to do herself. After introducing the themes with some video content they manage to draw the whole student group into a conversation inquiring into these themes, in which personal experiences are openly shared.

In his social studies classroom, high school teacher Ronald is talking to his students about the problem that an increasing percentage of elderly people in European populations is beginning to weigh heavily on national health care systems. One of the responses to this problem is an increase in informal caregiving, and as it happens a few students take the initiative to share that they are giving informal care to one of their parents. After this, Ronald takes on the role of conversation moderator and students start to share some experiences. At some point, the conversation turns to the question of whether a child should carry the responsibility of giving informal care to a parent.

As the cultural backgrounds of students in the classroom are diverse, and different cultures value informal caregiving differently, this quickly leads to a conversation about cultural differences in which students take diverse stances.

From this pattern of insight, so I wish to argue, an improved understanding of the relationship of students to complex societal challenges emerges, for having critically evaluated readiness-dichotomies, this insight leads to a similar evaluation of what we could call a space-dichotomy, meaning: a separation between the student-in-school and contemporary complex societal challenges "out there". Allow me to elaborate.

Given how most education is organized in today's world, it is quite intuitive to come to an understanding of educational space as separated from overall society, or the world "out there". Physically, schools do indeed typically appear as such spaces, in that they are buildings equipped with borders, in the form of walls, doors, and fences that enclose the school space. These borders, then, enclose what Arendt (1961) referred to as the educational realm. By understanding the world as "out there", the educator can think of his/her task as preparing students for the world and providing them with opportunities to participate in it. Yet these exemplary experiences and reflections of co-researchers suggest that students are not only participants-through-education, as through their evolving biographies they are involved in contemporary societal challenges regardless of education. Social anthropologist Tim Ingold (2008) argues that to think in terms of space-dichotomies is, in fact, typical for modern humanity as we tend to use a logic of inversion; we tend 'to turn the pathways along which life is lived into boundaries within which life is contained' (p. 1–2). From this point of view, it is interesting to note that Biesta (2013a), although inspired by her work, argues that the dichotomy Arendt imposed between the educational realm and the political realm is an unrealistic one, stating that 'it is not only irresponsible to try to keep political existence away from the school; it is also impossible to do so, because the lives of children

and young people – inside and outside the school – are permeated by questions about togetherness-in-plurality' (2013a, p. 118). Similarly, Dasberg (1975) argued that the separation between Youthland and the adult world is in many ways an illusion, as the kind of themes we typically associate with the adult world, such as sexuality and responsibility, are not fundamentally alien to the young. Following this logic, we might observe, for instance, that just as we need to move toward a more sustainable lifestyle as a society, so too the school building, the materials we use in the educational process, and the lifestyles of students and teachers are more or less sustainable. Or, that just as issues of inequality, poverty, polarization, and discrimination permeate society, so too the facilities and support systems are better in some schools than others, many schools struggle with bullying and social exclusion, and student populations are diverse in terms of cultural background, gender, sexual orientation, wealth, and so forth.

The complex challenges out there in the world are, then, also in here in school, and students are personally part of these challenges through their unique, evolving biographies. In complexity thinking, this phenomenon is often described as a fractal pattern (Davis and Sumara 2006); if we zoom in on a part of a whole (e.g. the school within society) we come to see the same patterns that we see when we look at the whole. Complex challenges, in other words, are not contained within closed spaces, but rather weave through the nested structures that organize our lives. Aware of these issues, Ingold (2008) poses the question: what happens if we reverse the logic of inversion? Then, he argues, instead of objects we will see lines of generation, and instead of insides and outsides (and thus boundaries), we will discern comings and goings. Notice, for instance, the way Ingold observes a tree (2010, p. 4):

> There it is, rooted in the earth, trunk rising up, branches splayed out, swaying in the wind, with or without buds or leaves, depending on the season. Is the tree, then, an object? (…) Where does the tree end and the rest of the world begin? (…) If I break off a piece in my hand and observe it closely, I will

doubtless find that it is inhabited by a great many tiny creatures that have burrowed beneath it and made their homes there. Are they part of the tree? And what of the algae that grow on the outer surfaces of the trunk or the lichens that hang from the branches? Moreover, if we have decided that bark-boring insects belong as much to the tree as does the bark itself, then there seems no particular reason to exclude its other inhabitants, including the bird that builds its nest there or the squirrel for whom it offers a labyrinth of ladders and springboards. If we consider, too, that the character of this particular tree lies just as much in the way it responds to the currents of wind, in the swaying of its branches and the rustling of its leaves, then we might wonder whether the tree can be anything other than a tree-in-the-air. (…) These considerations lead me to conclude that the tree is not an object at all, but a certain gathering together of the threads of life.

To summarize his observation of the tree, as an exemplary organism, Ingold (2010) suggests that it lives in a state of *entanglement*; rather than being an entity that is turned into itself, it lives in the open as a gathering together of "the threads of life". The tree, for Ingold, is both itself an entanglement (i.e. the bark-boring insects, the algae, the bird, the squirrel, the wind, the soil, and so forth co-constitute each other) and an entity entangled in the wider world (i.e. the tree as part of a larger ecosystem). Therefore, Ingold prefers not to speak of environments that surround organisms, but of a relational field, of zones of entanglement, or meshworks (rather than networks) of interwoven lines. Students, in this view, cannot be isolated from the world "out there", nor are they surrounded by it. Rather, in their growth and movement, 'they contribute to its ever-evolving weave' (Ingold 2008, p. 8) and 'join in the processes of formation' (Ingold 2010, p. 5–6).

The premise of entanglement emphasizes, thus, how our being in the world is relationally mediated (i.e. we fundamentally exist and develop through relationships), and thus implies a relational ontology. To explicate this point further I turn, now, to Karen Barad. Resonating with Ingold's analysis of the logic of inversion, and strongly leaning on the quantum physics of Bohr,

Barad develops the understanding that as human beings we excel in performing so-called agential cuts (2007): we consider different entities and their agency in isolation yet in doing so analytically cut the relational threads that our existence depends on. The primary ontological units, so she proposes, 'are not "things" but phenomena – dynamic topological reconfigurings/entanglements/relationalities/(re)articulations of the world' (p. 141). This is not to say, so is to be clear, that identities, both individual and collective, are unimportant or illusionary, but rather that these do not exist outside of or prior to their dynamic entanglement in the differential becoming of the world. In this view, it is in relationality that the experiences of self and other emerge (Barad 2007; Ceder 2019). The two fundamental experiences of being an entangled self are, so it then follows, (1) to be a whole with a hand in shaping the world's becoming and (2) to be a part shaped in the world's becoming (see also, Laszlo 1996).

Both Ingold and Barad do not hesitate to observe that an overly rigid attachment to the tendencies to inverse paths into spaces and to cut relationality into separation falls hopelessly short in the face of complex challenges confronting us, and they find other complexity thinkers I have been engaged with so far on their side. Crowell and Reid-Marr (2013, p. 15), quoting Einstein (in Nadeau and Kafatos 2001, p. 179), go so far as to frame the mindset of separation, as 'the biggest delusion of our time'. Wahl (2016, p. 83), building forth on the analysis of physicist Fritjof Capra, speaks of 'a crisis of perception', arguing that the failure to see 'the "hidden connections" that maintain the long-term viability of life as a whole' underlies the ecological, environmental, social and economic crises we are facing today. Understanding students' relationship to complex societal challenges in terms of entanglement is, thus, deeply complexivistic. Not only does it recognize that once we inquire into such challenges we encounter dynamism, nestedness, and interconnectedness, but it emphasizes furthermore that the students in our schools – indeed even the very young ones (Malone et al. 2020) – are not outsiders to all this, that situated within this interconnected world we are complicit in its ongoing formation (i.e. we are shapers) just as the development of our identities bears witness of the particular ways in which these challenges touch our lives (i.e. we are shaped). This understanding, notably, is a further specification of the perspective developed earlier in this chapter that we ought to approach students not merely as in-preparation-for-future-participation, but also as participants-here-and-now. It specifies, to be precise, that this participation-here-and-now is not just an opportunity provided by a teacher, as a gift bestowed on an otherwise disconnected individual, but rather, as we are relational beings in an interconnected world, an a priori given that can be explored, strengthened, and transformed through education. The premise of entanglement, so is the suggestive conclusion of this chapter, offers a powerful lens for developing a pedagogical response to complexity. Let us get at it.

References

Anderson, M. L. (2003). Embodied cognition: A field guide. *Artificial Intelligence,* 149, 91–130.

Arendt, H. (1961). The crisis in education. In H. Arendt (Ed.), *Between past and future: Six exercises in political thought* (p. 173–196). London: Faber and Faber.

Barad, K. (2007). *Meeting the universe halfway: Quantum physics and the entanglement of matter and meaning.* Durham: Duke University Press.

Biesta, G. J. J. (2010). *Good education in an age of measurement: Ethics, politics, democracy.* Boulder, CO: Paradigm Publishers.

Biesta, G. J. J. (2013a). *The beautiful risk of education.* Boulder, CO: Paradigm Publishers.

Biesta, G. J. J. (2013b). Responsive or responsible? Democratic education for the global networked society. *Policy Futures in Education,* 11(6), 733–744.

Biesta, G. (2020). Risking ourselves in education: Qualification, socialization, and subjectification revisited. *Educational Theory,* 70(1), 89–104.

Ceder, S. (2019). *Towards a posthuman theory of educational relationality.* New York, NY: Routledge.

Crowell, S., & Reid-Marr, D. (2013). *Emergent teaching: A path of creativity, significance, and transformation.* Lanham, MD: Rowman & Littlefield.

Dasberg, L. (1975). *Grootbrengen door kleinhouden als historisch verschijnsel.* Meppel: Uitgeverij Boom.

Davis, B., & Sumara, D. J. (2006). *Complexity and education: Inquiries into learning, teaching, and research.* New York, NY: Routledge.

Eriksen, K. (2006). The constructive developmental theory of Robert Kegan. *The Family Journal: Counseling and Therapy for Couples and Families,* 14(3), 290–298.

Freire, P. (1972). *Pedagogy of the oppressed.* Harmondsworth: Penguin.

Ingold, T. (2008). Bindings against boundaries: Entanglements of life in an open world. *Environment and Planning A: Economy and Space,* 40(8), 1–15.

Ingold, T. (2010). Bringing things to life: Creative entanglements in a world of materials. *ESRC National Centre for Research Methods,* Realities Working Paper 15.

Kegan, R. (2018). What "form" transforms?: A constructive-developmental approach to transformative learning. In Illeris, K. (Ed.), *Contemporary theories of learning* (2nd ed.) (p. 29–45). London: Routledge.

Kelchtermans, G. (1993). Getting the story, understanding the lives: From career stories to teachers' professional development. *Teaching and Teacher Education,* 9(5/6), 443–456.

Laszlo, E. (1996). *The systems view of the world: A holistic vision for our time.* Cresskill, NJ: Hampton Press.

Malone, K., Tesar, M., & Arndt, S. (2020). *Theorising posthuman childhood studies.* Singapore: Springer.

Meirieu, P. (2016). *Pedagogiek: De plicht om weerstand te bieden* (S. Verwer, Vert.). Culemborg: Uitgeverij Phronese

Nadeau, R., & Kafatos, M. (2001). *The Non-Local Universe: The New Physics and Matters of the Mind.* Oxford: Oxford University Press

Nussbaum, M. (2010). *Not for profit: Why democracy needs the humanities.* Princeton, NJ: Princeton University Press.

Olssen, M., & Peters, A. (2005). Neoliberalism, higher education and the knowledge economy: from the free market to knowledge capitalism. *Journal of Education Policy,* 20(3), 313–345.

Reid, A., Jensen, B. B., Nikel, J., & Simonovska, V. (Ed.) (2008). *Participation in learning: perspectives on education and the environment, health and sustainability.* New York, NY: Springer.

Rittel, H. W. J., & Webber, M. M. (1973). Dilemmas in a general theory of planning. *Policy Sciences,* 4, 155–169.

Rousseau, Jean-Jacques (1979). *Emile, or On Education* (A. Bloom, Trans.). New York, NY: Basic Books.

Scott, L. A. (2017). *21st Century skills early learning framework.* Partnership for 21st Century Learning, retrieved 24-09-2018 from http://www.p21.org/our-work/elf.

Shapiro, L., & Stolz, S. A. (2019). Embodied cognition and its significance for education. *Theory and Research in Education*, 17(1), 19–39.

Tijsma, G., Hilverda, F., Scheffelaar, A., Alders, S., Schoonmade, L., Blignaut, N., & Zweekhorst, M. (2020). Becoming productive 21st century citizens: A systematic review uncovering design principles for integrating community service learning into higher education courses. *Educational Research*, 62(4), 390–413.

Wahl, D. C. (2016). *Designing regenerative cultures.* Axminster: Triarchy Press.

In this chapter, I present the remainder of my collaboration with teachers as co-researchers. The initial inquiry presented in Chap. 2 generated 12 rich narratives in which various exemplary teaching experiences of co-researchers are described. These narratives have helped focus the inquiry on what I henceforth refer to as *pedagogy of entanglement*. The central argument behind this move has been that as long as our pedagogies approach the student first and foremost as someone to be prepared for and/or introduced into the world, we shall fail to illuminate and embrace the complexity of life and life's challenges. We shall fail, more precisely, to educate that part of the student that is touched by the challenges of world-in-motion and has a hand in their ongoing evolution in a here-and-now sense (i.e. we shall fail to have a pedagogical response to students' entangledness). Having come to this point, inquiry with co-researchers has to turn to the articulation of axiological and praxeological perspectives that can help teachers to engage with their students' entangledness meaningfully. Within the dimension of axiology, the question becomes: what is it that we, as teachers, wish to do to the entangledness of our students, and why? Within the dimension of praxeology, the question becomes: what can we, as teachers, do to help bring about those learning processes that we aspire to? I refer back to Chap. 1 to restate that I seek to contribute to answering these questions by generating interpretative, hermeneutic lenses through which teachers can inquire within the specific practices they are immersed in, so that situated insights and possibilities can emerge. The 12 biographical narratives hint toward numerous axiological and praxeological perspectives of this kind. How, then, are we (i.e. my co-researchers and I) to transform these initial biographical narratives into a set of helpful pedagogical perspectives?

A Diffractive Approach

To develop a proper strategy at this point, I considered it important to be aware that the inquiry with co-researchers can itself be understood as a process of learning in and from entanglement. After all, my co-researchers are teachers who resonate with the aims, teaching experiences, and considerations that drive my research (i.e. they are entangled with it in a particular way), and I attempt to enable a collaborative inquiry through which we can create perspectives that help not only other teachers but, also, ourselves. This insight – the insight that we are entangled researchers – is, in fact, an important methodological concern in complexity thinking (Davis and Sumara 2006). The entangled researcher theorizes not by looking at things from a distance but by 'dwelling in them' (Polanyi 2009, p. 18) and not by intervening from outside but by 'intra-acting, from within, and as part of, the phenomena produced' (Barad 2007, p. 56). This approach is a fundamental move away from how theorizing is often understood, namely as an external act of measurement and explanation (Barad 2007; Bozalek and Zembylas 2017; Gergen 2014; Haraway 1997). This move is, indeed, identical to the move from an ontology of individualism – 'the belief that the world is populated with individual things with their own independent sets of determinate properties' (Barad 2007, p. 19) – to a relational one.

From this point of view, there is an important remark to make about the status of the biographical narratives that the initial inquiry with co-researchers produced and, consequently, about what is needed in moving forward from this point onward. The biographical accounts of co-researchers are interpretations of past experiences. Apart from focusing co-researchers' attention on specific types of experiences, I have attempted to position myself as neutral as possible, encouraging co-researchers to find their own words for and interpretations of past experiences. I have encouraged them, in other words, to take a reflective step, to look back at the past and re-articulate it. In Chap. 1 I already stated that these biographical accounts cannot be understood as factual descriptions of what happened, but rather are to be seen as proof for how co-researchers interpret past experiences. I also argued that these interpretations are to be understood as co-constructions, for had I framed my questions differently other elements would have been foregrounded in co-researchers' narrations. I can now add, to these statements, an additional consideration: as co-researchers are themselves entangled in the phenomena of inquiry, their interpretation is unavoidably incomplete and dynamic (Barad 2007; Polanyi 2009); inevitably co-researchers are at every moment in time blind to certain other potentially helpful interpretations and once the particular way in which a co-researcher is entangled transforms so does his/her interpretation of past experiences. The biographical accounts illustrate, first and foremost, through what kind of lenses co-researchers ascribe meaning to their practice at the point in time and space of narration. The aim of this book is, however, not to identify these lenses as such,

but to articulate helpful perspectives that can inspire teachers. My reasoning has been, following these considerations, that my co-researchers and I are more likely to end up with rich, helpful perspectives if we challenge and transform initial biographical accounts than if we take them for granted and reflect the insights articulated in them. Reflection, to quote Haraway (1997, p. 16), 'only displaces the same elsewhere', and to reach our aims we should thus avoid this fallacy of representationalism (Barad 2007) and opt for a creative engagement with data through which – yet invisible – patterns of insight can emerge.

The more common, representationalist thing to do at this point of the inquiry would be to perform a content analysis (Denzin and Lincoln 2000; Elo and Kyngäs 2008), a procedure in which a coding scheme is derived from the data to describe the qualitatively different statements they contain. Such an approach, we can now see, would not serve my aims in this chapter. An alternative approach is needed to enable a creative process in which my co-researchers and I move and act with the data, so that insights can emerge that bring us a step further than merely representing what we already know. For this purpose, I have found inspiration in the notion of diffraction (Barad 2007; Bozalek and Zembylas 2017; Haraway 1997). The metaphor of diffraction contrasts that of reflection, and the difference can perhaps most easily be understood by explaining them as optical metaphors. The metaphor of reflection is illustrated by a mirror, which presents us with an image of what we position in front of it (e.g. performing a content analysis on qualitative data). The metaphor of diffraction, on the other hand, has everything to do with patterns of difference that emerge through material engagement, and refers to 'the way waves combine when they overlap and the apparent bending and spreading of waves that occurs when waves encounter an obstruction' (Barad 2007, p. 74). As Barad summarizes (p. 71): 'both are optical phenomena, but whereas the metaphor of reflection reflects the themes of mirroring and sameness, diffraction is marked by patterns of difference'. To utilize this metaphor in my collaboration with co-researchers, the diffractive methodology of Van de Putte et al. (2020) proved particularly useful to me. The interesting part of their methodology, for my current purposes, is that they too start with the biographical narration of past experiences and, from that point onward, develop an approach based on what they call diffractive scripts. Rather than static descriptions of the past – dead, in a sense –, diffractive scripts are dynamic explorations in the here-and-now – very much alive – that transform and gain meaning through a collaborative engagement of the researchers. The obstructions that facilitate the iterative development of diffractive scripts are constituted by the particular ways in which this collaborative engagement is intentionally organized, and the reasoning behind these choices needs to be transparent (for this, as well, an excellent example is provided by Van de Putte et al. 2020). In a general sense, these obstructions always invite reading one narrative through another to arrive at more creative insights (Bozalek and Zembylas 2017). In the remainder of this chapter, I present the steps through which the diffractive inquiry with co-researchers took shape. For each step, I describe how it was informed by previous steps, how it was designed as a diffractive tool, and which patterns of insight it produced.

Step 1.1: Creating Diffractive Scripts

Inspired by the work of Van de Putte et al. (2020), I designed a plan to transform the 12 initial biographical narratives of individual co-researchers into 3 diffractive scripts. Unfortunately, one co-researcher (i.e. Anika) decided to leave the inquiry, for finding herself in a particularly stressful period she considered participation too time-consuming. Proceeding with 11 co-researchers, I decided to bring them together in two groups of four and one group of three and to organize a session with each group. As I did not want to predict what the outcomes of the inquiry were going to be, I decided to form groups based on the availability of co-researchers and diversity in terms of school type (i.e. primary school, high school, applied university).

As these group sessions were to mark the start of a new phase of the inquiry, and for some co-researchers, a considerable amount of time had passed since their narrative-biographical interview session, I decided to be very concise in inviting them. Especially important, herein, was to present how our mutual inquiry, in its interweaving with literature study and autoethnography, had moved me to articulate the premise of entanglement as a hermeneutic lens for further inquiry. To create the opportunity for discussion on this matter, I organized a phone call of approximately 30 min with each co-researcher individually. In these conversations, I gave a transparent overview of how the three modes of inquiry had interwoven in the initial inquiry (as presented in Chap. 2). Also, I took a moment to explore, together, how students' entangledness is revealed in the biographical narrative that we had constructed together in the initial inquiry. These conversations were inspiring and provocative, and all teachers agreed that the premise of entanglement summarized and illuminated an important insight from the initial inquiry, and provides a promising lens for further inquiry. In these phone calls, I also provided an update of the steps I intended to take in the next phase of the inquiry. I described, in short, that I intended to facilitate a diffractive inquiry based on working with diffractive scripts. On several occasions, discussing this approach considerably helped me in fine-tuning the plan. I especially remember my talk with Ronald who, being experienced as a writer and director of theater performances, offered me valuable advice about how to provide the right amount of structure and freedom for collaborative script-writing. Following these phone calls, I sent an email to my co-researchers to ask for their availability for a group session. Attached to this email, I also sent them an informative letter, which once more summarized the process and outcome of the initial inquiry, and my plans for further inquiry. Based on co-researchers' availability, I then formed the three groups and scheduled the group sessions. The first group consisted of Pien, Ronald, and Sandra, the second group of Jacob, Astrid, Jens, and Kasper, and the third group of Aafke, Steven, Irene, and Elmarie.

Due to COVID-19 regulations, each group's session had to be organized as an online video call, using Microsoft Teams. In each group session, I first asked co-researchers to read each other's biographical narratives (i.e. of the other group members) and highlight so-called hotspots (Bozalek and Zembylas 2017), elements or sections that attract attention as they are, for instance, particularly provocative, disturbing, new, or recognizable. After this, I asked co-researchers to engage in a creative dialogue with each other based on a simple, yet in practice highly challenging, assignment:

> Formulate, as a group, one challenging yet realistic educational situation in which students' entangledness is brought to the foreground and, subsequently, develop and write down several scenes that describe how this situation develops in interaction with you as their teacher. In doing so, carry the narratives that you just read of each other with you as inspirational material.

For this assignment, I gave the group about 1.5 h at most. To develop a diffractive script online, we worked with Padlet, a tool that allowed us to work simultaneously in a mind-map-like online environment. I took on the role of facilitator, which in practice meant that I would if needed bring the group back to the assignment (i.e. develop scenes, write them down), and provide them with structure (e.g. with such a comment as 'let's slowly work toward the last scene of this script, let's say we have max 2 scenes left'). Also, co-researchers would naturally end up exploring and discussing various considerations and emotions that inform certain specific teacher decisions, and in such situations, I would encourage them to include such considerations into their scripts.

The diffractive strategies deployed in this assignment are: (1) starting with highlighting hotspots in each other's narratives, and encouraging co-researchers to utilize these in the process of script-writing, enables a creative reading of narratives through one another, (2) the assignment to formulate, first of all, a challenging yet realistic starting scene invites co-researchers to connect the script-writing process to what is currently hot in society and their recent personal teaching experiences, and (3) the challenge to

formulate scenes as a group, rather than individually, works as an obstruction, forcing co-researchers' perspectives to enter into creative dialogue. I now move on to present the three diffractive scripts that were created in these sessions one by one.

Diffractive Script 1 – The Multicultural Classroom – Pien, Ronald, and Sandra

Scene 1: A multicultural group of fourth-grade secondary school students in the Netherlands has an English class. During a presentation of a newspaper article about criminality, student A shouts out: 'I bet those f*cking Moroccans did it!'. The teacher doubts if A only means to joke, or if A is serious as well. Student A is from a non-Moroccan background, but there are Moroccan students present as well.

Scene 2: The other students remain passive, apart from some repressed laughter and some students looking furtively at each other. The Moroccan students in the classroom don't respond, as if they don't care. Students are waiting to see how the teacher will react. The teacher sees and feels that things are not quite right...

Scene 3: The teacher has a dilemma: continue the class as planned or engage with A's statement. The teacher knows that A's statement touches upon a recurrent theme. Also, the teacher feels personally uncomfortable, for as a teacher s/he wants to avoid discrimination and prejudice, but on the other hand, s/he also recognizes the thought that the criminal is likely Moroccan. The teacher decides, therefore, not to let the moment slip, and to try to open a conversation.

Scene 4: The teacher says the following to the students: 'I am not sure how you feel right now, but what A just said, as a joke or out of anger, affected me. It makes me feel uncomfortable and I feel that I cannot just continue as if nothing happened. Do you feel something similar?'

Scene 5: One student responds by saying: 'don't make it such a big deal, can't we just continue?'

Scene 6: The teacher responds that s/he indeed finds this a big deal and that that's why s/he wants to talk about it. And that s/he is curious if others experience this as a difficult topic as well.

Scene 7: The classroom becomes noisier and several students start talking to the teacher and each other. Several students express that they found A's comment stupid.

Scene 8: The teacher tries to ask some follow-up questions, to get more specific comments from students; 'what do you mean by stupid?', 'how does this affect you?', 'what would you like to say to A in response?'.

Scene 9: Meanwhile, the teacher notices that class ends in 5 min and that the conversation needs to be brought to a good end or needs to be continued later. The teacher says: 'this class is almost over so unfortunately, we cannot end this conversation today. I would like to continue this conversation though, and I notice that it triggered you as well. I'll think about how we can come back to it next time.'

Diffractive Script 2 – The Sustainable School – Jacob, Astrid, Jens, and Kasper

Scene 1: Two students in a secondary school in the Netherlands send a message to their physics teacher stating that they want to start a sustainability committee at the school. The students decided to seek the help of this teacher specifically for s/he is known to be strongly interested in sustainability. Although the students' question is not primarily focused on physics, the teacher wants to utilize their interest.

Scene 2: The teacher wants to speak to the students as soon as possible and, to have enough time for them, invites them to drop by at the end of the school day. In this meeting, the teacher starts by asking the students what their intentions are. They explain why they care about sustainability, and in the eyes of the teacher, they appear quite dedicated to their cause. The teacher shares his/her enthusiasm and, getting a bit carried away, together they

envision all kinds of initial ideas for a sustainability committee.

Scene 3: That night at home the teacher starts thinking about how to follow up on today's meeting, and what his/her role as a teacher can be. Multiple considerations enter his/her mind... 'Do I stay as close as possible to the interests and goals of these students and/or shall I connect it to certain specific educational goals? Should I try to connect this project to the learning objectives of physics? How do I attune myself to these specific students? I see that one of them has a lot of talent but lost his/her motivation for school whilst the other is more motivated in class, do I take this into account? How do I retain their enthusiasm? How can I stimulate a critical attitude in this project, around questions such as: what is sustainability exactly, and what are other perspectives than yours? To what degree do I foreground my values in my role as a teacher? Can the fact that I am pro-sustainability prevent the students from taking a critical stance and, on the other hand, can my enthusiasm encourage and energize them?'

Scene 4: The teacher decides to stay close to the enthusiasm of the students and the next day offers to help them set up a sustainability committee at school. The teacher does, however, formulate three intentions that s/he shares with the students... 'Let's try to make this interesting for as many students as possible and to include them, for instance by seeking opportunities to connect to the curricula and other students' interests. Also, let's try to get the school management on board, for which you should articulate and justify your plans. Lastly, let's act, let's make sure we make small concrete steps on short notice, whether successful or not, to get the ball rolling.'

Scene 5: It is 3 months later and it's the end of the academic year. A lot has happened over the past 3 months, including both experiences of success and frustration. To wrap up the initiative for this year, a meeting is organized for interested students and teachers. In this meeting, there is an open stage for sharing and performances. As part of this, the two students and the teacher challenge each other to each share, in a self-chosen way, some reflections around such questions as: did you notice any change in your interaction with teachers and students? Have you come to appreciate certain school subjects and their relation differently? What have you achieved? What went well and what do you want to learn or improve? What has surprised you and do you see certain things differently?

Diffractive Script 3 – Mock and Prejudice – Aafke, Steven, Irene, and Elmarie

Scene 1: A group of fifth-grade secondary school students at an International Boarding School in the Netherlands has a Dutch class in which speeches of various American presidents are watched. The students come from all over the world and about half of them live at the school dormitory. The students are in the middle of an assignment to analyze and compare political speeches. At some point, the teacher shows two mocking cartoons about (at that time) sitting president Donald Trump and expresses his/her disapproval of Trump. This triggers a student with an American background, who says to be a follower of Trump. The student does so politely, which is typical for an International Boarding School.

Scene 2: The student moves on to share his/her opinion and provides two arguments. Firstly, so s/he argues, America has historically made itself very dependent on international politics whilst job availability on the internal labor market is problematically low. Trump prioritizes this job availability. Secondly, the student suggests, Dutch media is very negatively biased toward Trump.

Scene 3: The teacher listens to the student and as a consequence feels confronted with his/her own Western European "bubble". S/he experiences a blind spot. This triggers all sorts of questions and considerations on behalf of the teacher... 'Do I create space to zoom in on this, or do I come back to it sometime later? If so, how much time and space do I create, and

for what exactly? Do I focus on content or process, emotion or reason? How does this affect the goal of this class?' In that moment, the teacher decides to create space to try to have a group interaction about having prejudices or blind spots, and about diversity in perspectives. The teacher is aware that students may not so easily be open and vulnerable about this topic.

Scene 4: The teacher decides to show vulnerability and acknowledges, in front of the students, that the student's comment makes him/her aware that s/he taught in line with his/her convictions. To trigger conversation, and allow students to think for themselves, the teacher gives them some time to individually reflect and write some thoughts down. To facilitate this, the teacher gives the following questions to the students: do you recognize the experience of having a certain blind-spot or one-sidedness in your thinking? Do you think this is a bad thing? What influences how you view things and how could you open yourself to other points of view? Do you have an example of an occasion in which you opened yourself to another view than your own?

Scene 5: Meanwhile, the teacher notices that there is only a little time left for interaction. S/he decides to follow up on this theme during the next class later the same week. S/he tells the students that in that class they will engage in dialogue in rotating pairs of two, and s/he asks to prepare by writing down some thoughts about a topic on which they have a strong, one-sided opinion.

Step 1.2: Mapping Initial Insights

To harvest pedagogical insights triggered by co-constructing a diffractive script, I followed up on the assignment of step 1.1 with a group discussion of approximately 1 h. Initially, my idea was to have this discussion directly after constructing the diffractive script, but as step 1.1 was both time- and energy-intensive – especially in an online environment – two of the three groups decided to delay this discussion to a later day the same week. My idea behind this harvesting process, was twofold: (1) to attempt, as a group, to recognize what kind of patterns of insight emerged from transforming the initial biographical narratives into one diffractive script, and (2) to map these insights in such a way that they could, in turn, be engaged with in subsequent diffractive steps aimed at bringing the inquiry yet further. Therefore, I asked co-researchers to not only share insights but to map them into a web of insights. To facilitate this, we kept working with Padlet, and the assignment was as follows:

> Share, with each other, what kind of insights regarding the question of how you can work meaningfully with the entangledness of students emerge from co-constructing the diffractive script and map these insights into a mind-map structure in Padlet.

At the beginning of this assignment, I proposed one important diffractive principle to the group, namely: tensions, for instance in the form of contradictions, paradoxes, or dilemmas, are allowed. The aim, namely, was not to come to some sort of agreement with each other in which all tensions are resolved, but rather to map emerging insights in their diversity and richness, and it is exactly at the point where tension emerges that further diffractive inquiry can be most fruitful.

My role in this harvesting process was, again, that of a facilitator. I illustrated how to create a mind-map structure in Padlet and I invited the group to assign someone to create the mind-map structure or to share this task. During the process, I tried to assure that shared insights indeed got written down in the mind-map, and on occasion helped to summarize insights harvested thus far. Sometimes, groups asked me to help write the insights down, as it was quite challenging for them to combine discussing insights and writing them down. On such occasions, I made sure to write things down in the words of co-researchers and always checked if I formulated it correctly, and if the insight was located at the proper location in the mind-map. Sometimes, the group would conclude that the mind-map needed to be somewhat reorganized, to group certain insights together, or connect them to other insights in a particular way. Furthermore, I asked facilitating questions such as 'does anyone have a different insight, which has not been shared thus far', or 'does anyone recognize, perhaps in a slightly different way, the insight shared by co-researcher X'?

After a harvesting session was finished, I always took a moment to make the mind-map a bit neater, by correcting grammar mistakes and visually clarifying the structure created (e.g. by dragging grouped insights closer to each other). After this, I sent both the diffractive script and the mind-map to the group by email and invited them to check if they felt that something accidentally got left out or framed wrongly, and to share this in a reply-to-all email. For each group, only a few small corrections were gathered through this communicative validation step, which I subsequently processed. Figures 3.1, 3.2, 3.3 present the three mappings of initial insights that were created through this process.

Step 2.1: Transforming Diffractive Scripts

The results of steps 1.1 and 1.2 provide three exemplary scripts of what pedagogy of entanglement might look like and rich mind-maps of associated pedagogical insights. Yet, there is still a lot that inquiry with co-researchers can contribute to my aims, especially since the mind-maps reveal certain tensions that invite further inquiry. Such tensions have to do, for instance, with the degree to which you, as a teacher, (1) share with students and/or keep to yourself how you are entangled with educational content, (2) follow students' initiative and/or resist them, or (3) follow curricular plans and/or deviate from them. There often appears to be a sort of balance to be found or a paradox to embrace, in the sense that in such tensions both sides have their value. It can well be argued that such paradoxes are what provide the educational process with creative potential, and should not be sought to overcome but rather to embrace (see Palmer 2017). Also, in reading the diffractive scripts you may get the feeling, as do I, that the scenes do not describe the only possible scenario, nor necessarily the best scenario. Rather, they are honest attempts,

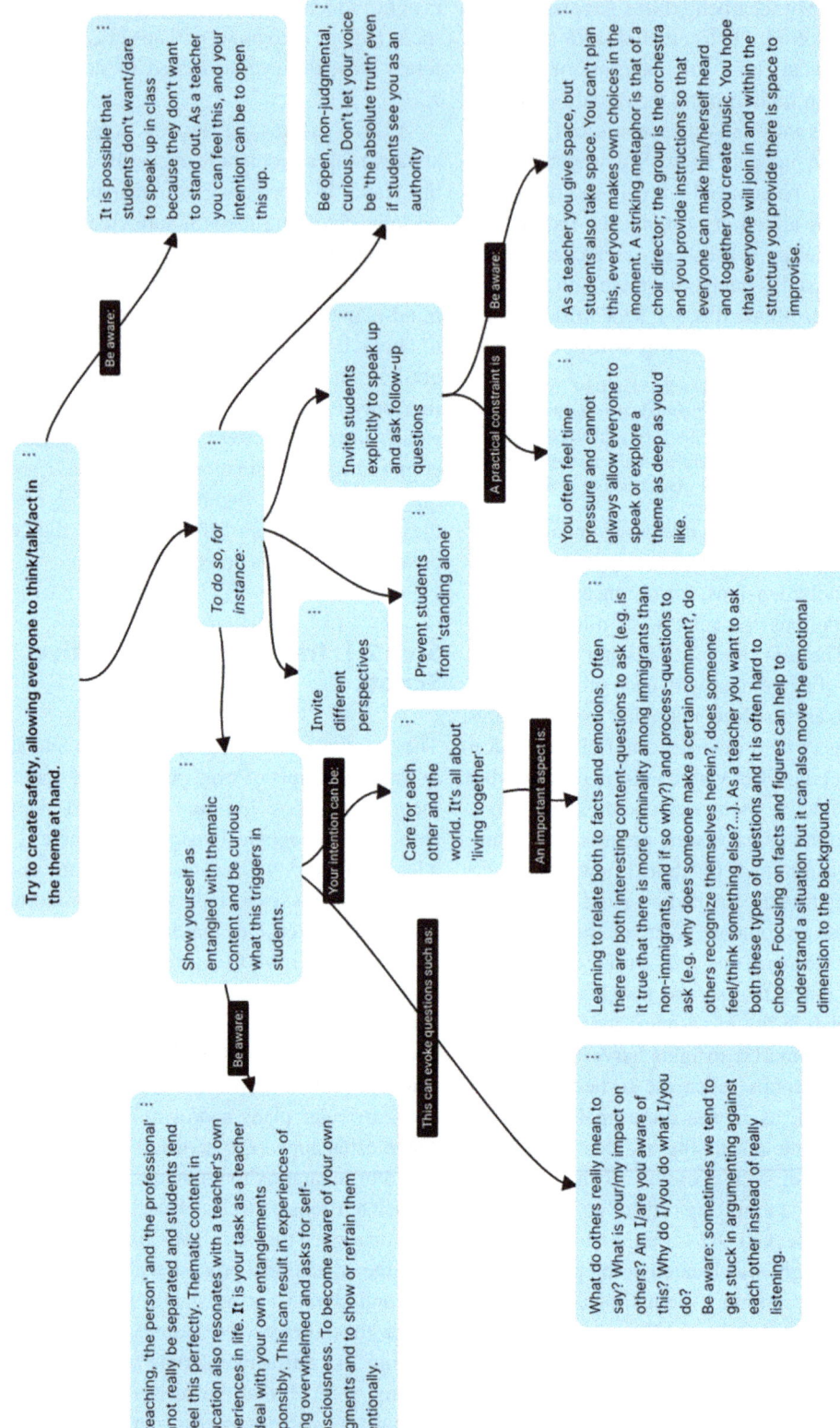

Fig. 3.1 Mapping of initial insights of group 1

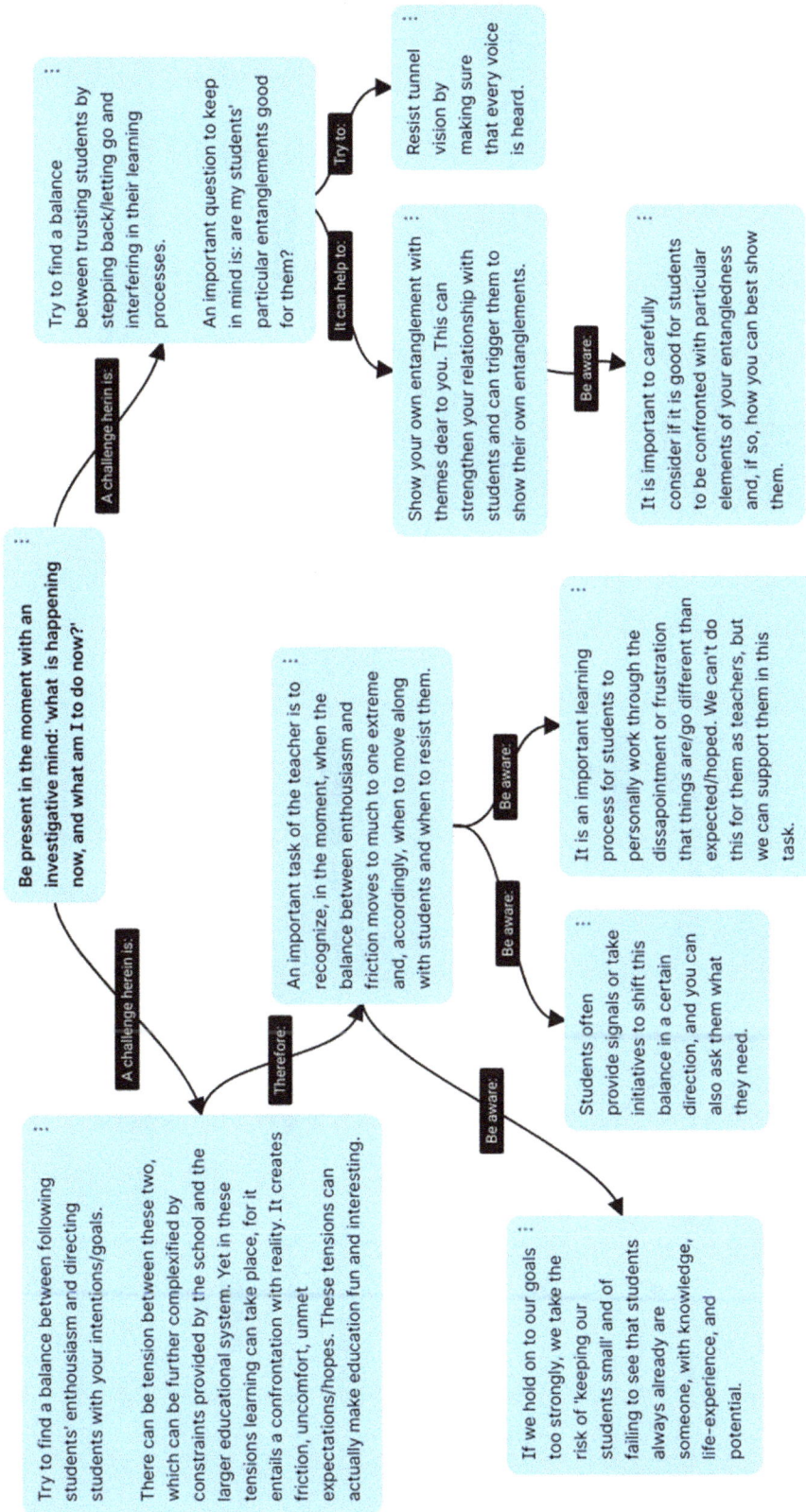

Fig. 3.2 Mapping of initial insights of group 2

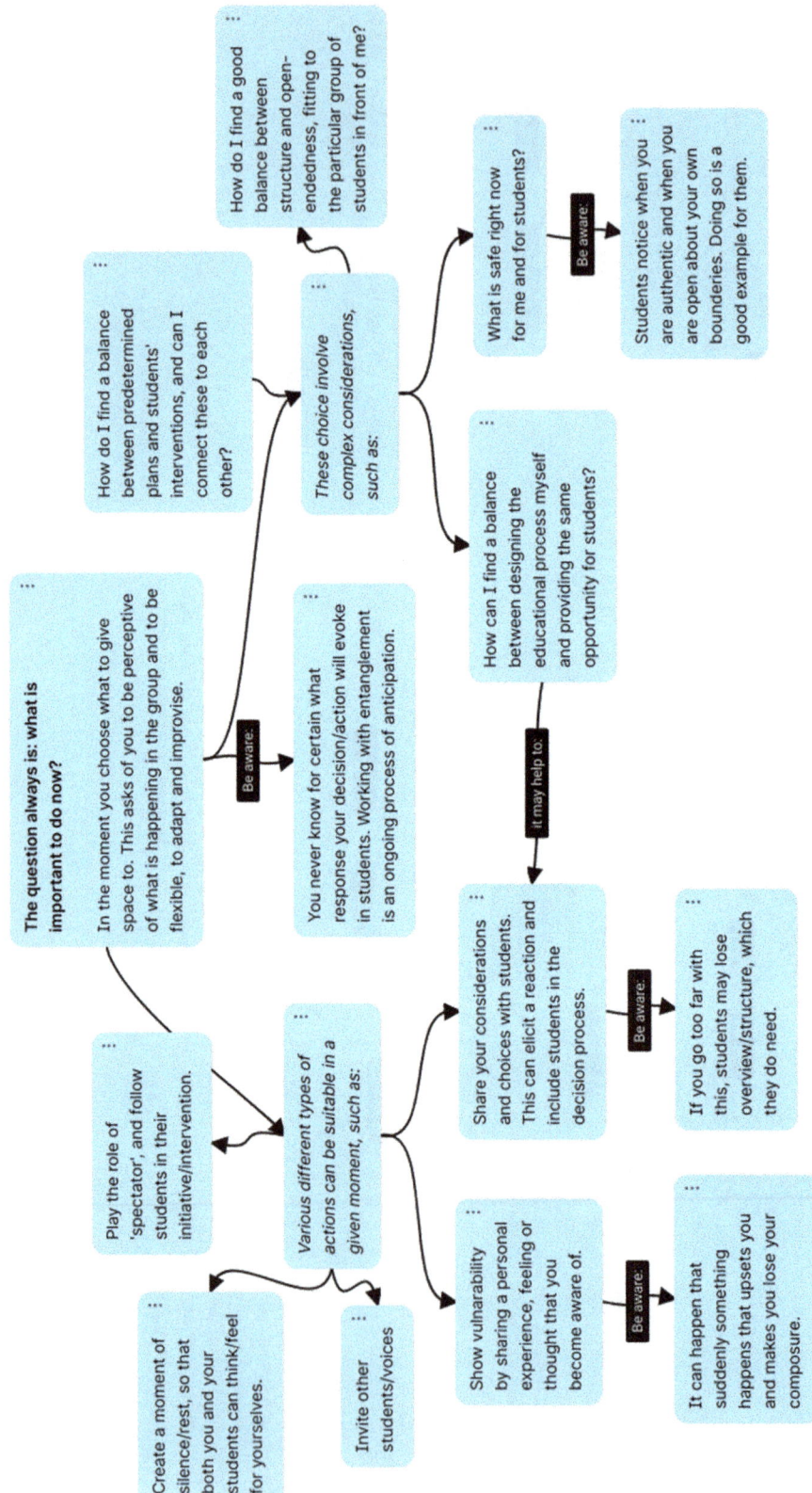

Fig. 3.3 Mapping of initial insights of group 3

informed by experience and insight, but inevitably imperfect and unfinished. It would be interesting, therefore, to imagine alternative turns that these diffractive scripts could take, and doing so can further illuminate the tensions that are at work. For these reasons, I decided to continue the inquiry with another diffractive step, which I shall again present in two sub-steps.

For this next phase of the inquiry, I decided to do another round of 1-on-1 research sessions with each co-researcher. I decided to meet them individually for two reasons. Most importantly, as co-constructing one coherent diffractive script is quite a challenging assignment, in which collaborative choices have to be made, I was curious to hear what co-researchers would change about the scripts if they were free to decide for themselves. As different co-researchers are likely to make different choices, this would likely illuminate certain tensions in the scripts. Also, now that patterns of pedagogical insight were starting to become clear, I was hoping to enrich them further by reconnecting to biographical teaching concerns and experiences of individual co-researchers. My intention at this point was, thus, to facilitate a re-reading of diffractive scripts through the narratives of individual co-researchers. To do so, I scheduled a 2-h session with each co-researcher. I was able to do 5 of these sessions offline, yet due to COVID-19 restrictions, 6 of them took place as an online video call through Microsoft Teams. In these sessions, I translated my intentions into several sub-assignments building on each other, in the following order:

1. Reread the diffractive scripts and scan the mind-maps of all three groups with the following questions in mind: what attracts your attention in particular, and are there certain things that you would do differently?
2. Now focus on the script of your group. Rereading it, can you tell me what attracts your attention and if there is something that you would like to suggest as an alternative or addition to the script?
3. Now focus on the script of one of the other two groups, can you tell me what attracts your attention in their script and if there is something that you would like to suggest as an alternative or addition to the script?
4. Looking at what has attracted you in rereading the scripts and mind-maps, and at the suggestions that you have made, can you identify a certain pedagogical tension (i.e. balancing, or choosing between, A and B) that is at play here and that attracts you particularly? How would you frame this tension?
5. If you extrapolate this particular tension to your own recent or current teaching experiences, which exemplary experience comes to your mind? Can you tell me about that experience?

Apart from posing these questions, I took on two roles in this phase of the inquiry. First of all, I tried to listen carefully and ask open questions, encouraging co-researchers to elaborate. Secondly, inspired by the work of Freire (1972) and by my experiences in steps 1.1 and 1.2, I took on the role of summarizer. That means: I repeatedly, and especially at the end of each sub-assignment, tried to summarize insights followed by the question: is that an accurate summary of insights, or did I miss something? In Freire's work, feeding such summaries back to co-learners is of crucial importance, as it enables them to become (more) aware of their current position and, if needed, move beyond it through critical engagement. I did, indeed, notice that once I provided such a summary to a co-researcher it would often trigger an additional insight or a further clarification. For my purposes, taking on this summarizing role furthermore assured that I would come to a mutual understanding with a co-researcher, rather than misinterpreting him or her. It assured, thus, communicative validity. As I recorded each session and made notes of summaries, this enabled me to provide an overview of outcomes. In what follows next, I present the alternative and additional choices that co-researchers suggested for the three diffractive scripts. Subsequently, I provide an overview of the tensions co-researchers zoomed in on and the resonating exemplary experiences they narrated. It is worth mentioning, that the experiences narrated by Astrid and Sandra are re-readings of earlier experiences that were also narrated in the narrative biographical interview conducted for the initial inquiry presented in Chap. 2. For all other co-researchers, the narrated experiences refer to experiences that they had after the initial narrative biographical interview session.

Additions and Alternatives to Diffractive Script 1 – The Multicultural Classroom

Extra steps in-between scenes 3 and 4:
- The teacher first asks the student: 'are you serious or are you making a joke?'
- The teacher first focuses group conversation on the facts: 'what are the crime statistics of the Netherlands?'
- Before having a group conversation the teacher first talks to the student 1-to-1: 'do you have certain personal experiences that made you say this? If not, what else made you say this?'
- Before possibly entering into group discussion, the teacher first draws a clear line, stating that s/he does not accept such a comment: 'you are entitled to your own opinions, but you can't say it this way, for that does not belong to the kind of open, respectful discussion that we strive for here.'

Addition to scene 4:
- The teacher tries to be very specific in sharing that s/he feels affected, so that this contributes to the content of the discussion: 'I have a close group of friends with three Moroccan guys in it, and we often talk about this topic. They told me how striking they find it that whenever such a comment is made, no one stands up to say that he or she finds this comment not OK.'

Alternatives to scene 4:
- The teacher gives a provocative response, saying 'I bet you're one of those rich kids who gets everything from daddy'. This results in a sort of "huh, what did the teacher just say- moment", which takes the edge of the situation and creates space for a

conversation. The teacher tries to use the momentum to collect more prejudices from the group and to explore, together, where prejudices come from and how we experience them.
- The teacher decides to facilitate discussion and to take a clear lead so that it doesn't become too messy/vague. S/he says: 'I can imagine that this comment is funny, but also that some of you feel affected. I want to ask you to spend 5 min in small groups to discuss if you find it acceptable to say such things and why. After that, we'll share opinions in the large group.' The teacher creates the groups randomly by giving numbers to students and walks around to help subgroups if needed.
- The teacher decides not to enter into discussion with the group at this moment, but mentions, instead, that s/he intends to come back to it on a later occasion.

Addition to scene 6:
- The teacher tries to be very specific in explaining why s/he finds this a big deal and wants to talk about it.

Additions and Alternatives to Diffractive Script 2 – The Sustainable School

Additions to scene 2:
- Before meeting the students, the teacher thinks about what his/her role and intentions can be in this initiative, and how much time and energy s/he wants to invest.
- The teacher openly discusses with the students where they can best pursue their interest in sustainability, and in doing so also asks for options outside of school.
- The teacher and students associate about different scenarios for how this initiative can potentially develop, and try to create an image of the path the students want to pursue.

Additional consideration and decision in scene 3:
- 'How can I use this initiative to help the student who lost his/her motivation to have experiences of success, and how

can I connect this to the school subjects s/he is currently struggling with?' The teacher decides to have a 1-to-1 conversation with this student about this, and if the student agrees, to talk with some of his/her teachers to see if certain curricular content can be connected to this initiative.

Additions to scene 4:
– The teacher challenges the students to gather as much knowledge/theory about sustainability as they can, both from sustainability advocates and sustainability adversaries. The teacher intends to have a critical conversation with the students after they do this, asking such questions as 'how do you know this is true?' and 'what are your sources?'.
– The teacher decides to have a conversation with the students about the question of what they would do in the case of encountering someone who denies climate change.
– The teacher formulates an additional intention: let's try to get a diverse group of students into the committee, representative of the different perspectives on sustainability within the school.
– The teacher weakens his/her intention about getting the school management on board, by explaining that it is not something that has to be done necessarily or immediately in his/her opinion, but that if they really want to achieve something in the school it is important to take into consideration.

Addition to scene 5:
– The teacher takes the initiative to invite potentially interested people from the local community and his network to the meeting so that they can watch students' presentations and meet each other.

Additions and Alternatives to Diffractive Script 3 – Mock and Prejudice

Alternative preparation before scene 1:

– In preparing for this class, the teacher visualizes how students might react to the mocking cartoons and based on that decides to show more diverse cartoons, for instance one of Trump and one of Obama. S/he also decides to refrain from sharing his/her own opinion of Trump, unless s/he finds it needed to "tickle" the students a bit, to elicit a reaction/discussion.

Extra step in between scenes 3 and 4:
– The teacher takes a moment to acknowledge that there are different points of view on Trump and that it is very valuable when someone shines a light on another point of view so that we can understand why such a big group of people feels unseen/unheard.

Alternatives to scene 4:
– The teacher asks a follow-up question to the student: 'how do you feel if something valuable to you is degraded?'. The teacher decides to continue on this topic in the next class, and later on designs a class circling the question 'what is holy to you?'
– The teacher decides to enter into dialogue with the students about the question 'do you think I, as your teacher, am entitled to have such a conviction, and if so how?' The teacher first asks students to talk about this question in duos and then facilitates a group discussion. The teacher also shares his/her position: 'I may have a personal conviction as a teacher, but I should always provide enough space for other points of view'.
– The teacher decides to stay as close as possible to the disagreement at hand, emphasizes how cool it is that we found a topic on which we don't all share the same opinion, and then zooms in by asking: 'let's hear it, what are the different opinions amongst us?'.
– The teacher decides to focus the conversation on the function of mocking cartoons, and on what it takes to accept or embrace mock even if it affects your own beliefs.

Astrid – The Tension Between Stepping Back and Interfering

Astrid is part of a group of teachers that wants to make it easier for students to make the transition from primary school to secondary school. She developed the idea that it could help to let fourth-grade students be buddies for new students, and that this could be an interesting project for her current third-grade students. She decided to share her plan with them and proposed that they could be buddies next year. Students reacted with enthusiasm. From this point onward, Astrid decides to step back and support students from the sideline. She says to her students: 'if you're ready with chapter 4 I'll give you my keys so that you can go to a separate room to develop a plan'. Astrid resists the urge to go to these students and direct them, thinking to herself: 'if I want them to take responsibility for this project, and not just do it for me, I have to step back and trust them'. Students come up with first plans and are currently developing them further with a great deal of motivation. Astrid keeps herself somewhat at the sideline, only occasionally stepping in to brainstorm with students, give them suggestions, and, especially, ask them questions to help them think for themselves.

Aafke – The Tension Between Zooming Out and Zooming In

Aafkes Fashion-academy students joined a seminar about racism. Afterward, in class, one of the students reacts rather emotionally. At the seminar, she had felt excluded as a white woman due to a response she had received when she asked a question to a panel. As she shows her emotion in class, Aafke sees other students "freeze". Aafke decides to zoom in on the incident and asks the student what happened exactly. After she describes it, Aafke also asks other students how they'd experienced the incident. Other students also found the reaction of the panel aggressive, and also felt excluded. At this point, Aafke suspects that the students are not fully aware of the systemic level on which racism had been discussed in this seminar. Therefore, she decides to zoom out, to situate the seminar in its context and facilitate a conversation about 'what's going on in terms of racism in the world right now?'. In this conversation, different students, with differ-

ent ethnic backgrounds, share their knowledge and experience. Aafke also shares her experience with racism as a Dutch person from an Indonesian background. After this, Aafke decides to zoom in on the incident in the seminar again, and asks the student: 'if you look at what happened again now, what do you see?'. The student's initially strong emotions have faded away, and she is now able to look at what happened from different perspectives.

Elmarie – The Tension Between Following a Student's Initiative and Sticking to the Plan

At the beginning of Dutch class, one of Elmarie's students says out loud 'I hope we'll have a talking class today'. Elmarie would like to respond to this invitation, but Elmarie decides to inhibit herself. She remembers that this particular student failed to finish reading a book for his aural exam last week, and thinks to herself 'I'm not gonna give him what he wants, this is gonna be my class'. Forty minutes later, after Elmarie finishes her prepared plans for this class and students have worked well, there are 10 min left and an off-topic group conversation starts developing. At that point, Elmarie makes contact with the student, and says to him 'aah, now we'll still have the talking class you hoped for!'

Irene – The Tension Between Moderating and Spectating

In Irene's Life-Orientation class, students start asking a lot of questions to a very religious student and in Irene's eyes, these questions appear quite offensive. Irene is ready to intervene, to moderate the interaction, but she notices that the student stays very calm and can manage the conversation and express herself clearly. Irene decides to remain a spectator to the interaction between students and afterward has a short chat with the student about how she experienced it.

Jacob – The Tension Between Showing and Concealing Your Own Entanglement

In his physics class, Jacob has arrived at the chapter on astronomy. Astronomy was one of the topics that Jacob did not choose to focus on when he was studying physics, for he was more interested in other topics. Yet, if he thinks about it now, he can identify certain questions about astronomy that fascinate him, like: 'what is the Sun made of?', 'how long will the Earth live?', and 'how do we know these things?'. How cool, Jacob thinks to himself, that we can determine what the sun is made of by looking at it from afar! Jacob realizes that the chapter is not really organized around such questions, yet that they do cover its content. To make matters more interesting for himself and his students, Jacob decides to share his fascination for these specific questions and organize the chapter around exploring them together. He conceals, however, that astronomy is a topic he has been less interested in throughout his studies and career.

Jens – The Tension Between Prioritizing Students' Enthusiasm and Conforming to Priorities of the School/System

Jens supervises law students in their bachelor end-projects. In supervising and grading students' performance, Jens has to use a rubric, but he finds this rubric very limited. In his opinion, it doesn't quite measure what he finds important and he seriously questions the very possibility of an objective grading procedure. Furthermore, he experiences that a strong focus on the rubric in the supervision process tends to decrease both his own and his students' motivation. Jens has decided, therefore, to let the rubric come second, and to prioritize students' enthusiasm. After an end-project is done, he determines based on his intuition and experience which grade he deems appropriate, and he then fills the rubric in such a way that it ends up with the grade he has in mind.

Kasper – The Attempt to Find the Right Degree of Opening Yourself Up to Your Students

In his microbiology course, Kasper starts every lecture with what he calls 'microbiology is everywhere'. He shares a personal fascination or something he experienced or saw that week with his students. His most recent sharings are a fragment from a South Park episode about the COVID-19 pandemic, some pictures from mushrooms he found on a hike in the forest, pictures of himself

and his children joining a vaccination campaign, and a video on the importance of vaccination. Kasper experiences that "microbiology is everywhere" creates a connection between himself and his students, and to microbiology as the subject of study. Sometimes, for instance, a student he crosses paths with in the hallway shouts out something like 'Hi Kasper, microbiology is everywhere!'

Pien – The Tension Between Creating Space to Inquire into a Certain Entanglement and Moving On

Next to her work in a primary school, Pien works 4-h a week for a so-called rebound center, where students come who cannot be in their secondary school for a while due to problems/conflicts. The goal is to help students to become able to go back to their school or to find them a more suitable school. Pien has a meeting scheduled with a student who is in the last phase of secondary school. She has an autistic background, has bad grades, and has serious problems at home. The aim is that she will do half of her exams this year and finish the remaining subjects next year. She enters Pien's office, Pien says 'good morning', and she looks at Pien with a depressed face. Pien says 'isn't it a good morning?'. She says 'well, I'm still alive'. Pien's intention for this meeting is to make a plan for the upcoming period so that the student will be able to do her exams as intended. Yet, at this moment she feels that she cannot just move on to doing so, and instead decides to inquire further. Pien decides to create the space and time to talk with the student about how she feels and to show her support. At some point, the student says 'I think I'll only do my English exam this year, and delay the rest to next year'. Pien takes this moment as her signal to move on to the matter of making a plan, and says to the student 'that's not what we're gonna do here, it's four subjects, together we can do it. Come, let's work on a plan!'.

Ronald – The Tension Between Not-Knowing/ Vulnerability and Control

Ronald, currently working as a teacher educator, receives an email from one of his students about her internship at a secondary school. Due to COVID-19, her internship includes a lot of online teaching, and she is frustrated with her performance and seeks Ronald's help. Instead of just giving suggestions for a solution, saying 'you can do this or that', Ronald decides to share his own frustration and struggle. He says: 'I also struggle... Sometimes teaching online really makes me unhappy and I find it very difficult, I am also still learning...'. The student replies that she really appreciates Ronald's answer and that it helps her to keep going.

Steven – The Tension Between Encouraging Students' Self-Expression and Being Critical

Steven's students are in the middle of an art project. One of his students comes to him for support. With strong conviction, she shares her belief in certain conspiracy theories with Steven, such as 'the elite drinks the blood of children', and asks Steven for advice on how to manifest this in her artwork. Steven feels overwhelmed, and finds himself caught up in an internal struggle; 'do I do something with the fact that she bases herself on information that I find highly questionable, or do I just help her to manifest a belief that she is entitled to have?'. At that moment, Steven decides not to criticize her, for he doesn't want to hurt her feelings, but later on a feeling of regret slowly grows on him. He wishes he would have said something like 'girl, where do you get these ideas?!'.

Sandra – The Attempt to Find the Right Degree of Showing Vulnerability to Your Students

In a marketing course that Sandra is teaching, she experiences a struggle. There is a new teaching method that she is expected to follow, but she feels uncomfortable with the method and notices that her students have issues with it as well. She decides, therefore, to share her discomfort and to use it as an opening for a group conversation around the question: 'how can we fine-tune this course together?'. Her question opens up a lively conversation, and together they fine-tune the course along the way.

Step 2.2: Patterns of Insight

After finishing step 2.1, the need arised to harvest insights once more. My aim, this time, was to recognize which patterns of insight emerged from rereading diffractive scripts through the narratives of individual co-researchers. Therefore, I collected insights 1-to-1 with each co-researcher in the remainder of the session described in step 2.1. To do so, I gave every co-researcher two additional sub-assignments, being:

6. If you look at how this creative tension is at work in the diffractive script, and how it resonates in your own teaching experience, what do you now realize about what it takes to work with this tension?
7. Now focus on the script of your group again, and connect with the opening scene it begins with. Having developed and altered a script departing from this scene, and having explicated and confronted creative tensions herein, how would you now summarize the pedagogical goal or intention that drives the teacher's efforts?

In this part of the inquiry, I took on the same role as in step 2.1 (i.e. asking open questions, and summarizing and communicatively validating insights). It was, for me, interesting to notice how on the one hand every teacher emphasized certain insights and formulated them in their own words, whilst on the other hand very similar patterns of insight emerged across all sessions. After finishing this procedure with every co-researcher, I, therefore, decided to attempt to make and share with all co-researchers one summary of insights, in which these patterns are emphasized and every individual insight is included. I shared this summary with all co-researchers in the form of a recorded PowerPoint presentation, and – for purposes of communicative validity – asked them to share with me if they missed something in the summary. I received, through this procedure, a few small corrections, which I then added to the summary. I now move on to summarize the praxeological insights harvested concerning what co-researchers realized about what it takes to work with a specific pedagogical tension. Subsequently, I present a summary of the axiological insights harvested per diffractive script.

What It Takes to Work with Tensions that Emerge in Pedagogy of Entanglement

Perceptiveness: trying to perceive (also referred to as sense, scan, signal, feel, read and notice), both in preparing for a class and during a class itself, what is happening in the moment. Especially:

1. What do you perceive in individual students?

   What kind of attitudes do students have?

   Do I notice or suspect any strong emotions or convictions concerning the themes we are working on?

2. What do you perceive in the group as a whole?

   How much safety and trust is there?

   Which tensions or differences are there?

   How 'hot' are the themes we are working on for the group?

3. What do you perceive in yourself?

   How do I feel right now and what are my impulses?

   What is my own relation to the themes we are working on?

   Where lies my passion and strength in this moment?

4. What do you perceive in society?

   How are the themes we're working on reflected by ongoing dynamics in society?

   Which different stakes and parties are involved?

5. Which curricular constraints and opportunities do you perceive?

   How much time and space do we have right now?

   Is there a deadline upon us?

   Is there a link with another subject/project?

Commitment: trying to have a clear, embodied pedagogical goal/intention and trying to be congruent in basing your actions hereon.

Authentic style: trying to stay close to your own passions and to ways of working that work for you, as also experienced by your students.

Professional realism: trying not to cross your own limits (e.g. emotionally, physically, idealistically) and to put your role into a realistic perspective which neither underestimates what you can give to students due to your life experience nor overestimates the reach of your influence, knowledge, and responsibility.

Constructive *self-awareness:* trying to be aware of your own entangledness with thematics at hand and the interventions you tend to enact as a matter of habit or reflex and to utilize this awareness in a way that serves the educational process.

Collegial support: trying to keep critically questioning and developing your pedagogical views and actions together with colleagues and other educational professionals.

Professional *independence:* trying to nurture the courage, calmth, and flexibility to make and justify your own choices in the moment, even if these diverge from initial plans and expectations.

Summary of Axiological Insights for Diffractive Script 1 – The Multicultural Classroom

By Pien, Ronald, and Sandra

We aim for:

Awareness of:
- what makes someone say what s/he says.
- the effect of stereotyping statements on ourselves, others, and society as a whole, and our role and responsibility in this.
- the insight that we all belong in this world and deserve to be included in society, to exist and live together.
- the necessity of certain rules, norms, and values about how we communicate with each other.
- the insight that freedom of speech goes hand in hand with the responsibility to

speak considerately and that you are responsible for the choices you make.

Safety and openness, that is:
- the openness to share what we feel, think and believe with each other, and the safety of doing so peacefully, in a humane way that does not exclude people but allows everyone to exist and live together.

Constructive action, that is:
- the practice to make your own, humane choices, and to take responsibility for your actions.
- the feeling and experience that you can solve a conflict and deal with challenges/ differences constructively.

Summary of Axiological Insights for Diffractive Script 2 – The Sustainable School

By Jacob, Astrid, Jens, and Kasper

We aim for:
Awareness of:
- the importance and complexity of sustainability, of how everything seems to flow into each other and keeps changing.
- the trends and bubbles in society, and their relation to each other.
- the distinction between opinions, facts, and fake news.
- what you think about sustainability, why you think that, and the shortcomings in your own thinking.
- the role you play with regard to sustainability and the role you want to play.

Respect, openness, and critical thinking, that is:
- the respect and openness for differences, for other viewpoints, yet without losing the capacity to criticize viewpoints and to resist the equation of fake news and science.

Co-shaping society, that is:
- learning to express yourself; forming your own critical opinion and being able to argue for it qualitatively.
- exploring and deepening your interests and enthusiasm, and turning this into action.

- the attempt to transcend "bubbles of sameness" and to grow toward more realistic and sustainable perspectives together.

Summary of Axiological Insights for Diffractive Script 3 – Mock and Prejudice

By Aafke, Steven, Irene, and Elmarie

We aim for:
Awareness of:
- what is holy/valuable for you, and how it affects you if this is threatened. Of how you identify yourself, where your identifications "solidified", and if you might perhaps benefit from "loosening" certain identifications.
- what is holy/valuable for others and why large groups of Trump supporters feel unseen/unheard.
- how manipulation/framing works (e.g. social media, news, cartoons) and what the "message behind the message" is.
- how your thinking and political beliefs are co-shaped by the environment you grew up in. That as a human being you are always part of groups and that personal identity and group identity are interwoven.
- what your actions/statements trigger in others?

Living together in a group, that is:
- being able to express your feelings, experiences, and thoughts. Being able to stand for something based on your background/ authenticity.
- striving toward "light/positive" entanglements with each other rather than "dark/ negative" ones.
- learning to pause and stand still for a moment.
- learning to listen to and respect what others have to say, to relate to their position and value their otherness, and to simultaneously continue to critically question both yourself and others.

Moving Forward

Through diffractive inquiry with co-researchers, as presented in this chapter, helpful axiological and praxeological perspectives for a pedagogy of entanglement have started emerging. To develop these further, I need to attempt integration with my other modes of inquiry, that is to say: I need to bring both axiological and praxeological outcomes into a generative conversation with state-of-the-art complexity thinking and with triggering experiences of my own within the process of inquiry. I take on this task in the three chapters that follow next.

References

Barad, K. (2007). *Meeting the universe halfway: Quantum physics and the entanglement of matter and meaning*. Durham: Duke University Press.

Bozalek, V, & Zembylas, M. (2017). Diffraction or reflection? Sketching the contours of two methodologies in educational research. *International Journal of Qualitative Studies in Education, 30*(2), 111–127.

Davis, B., & Sumara, D. J. (2006). *Complexity and education: Inquiries into learning, teaching, and research*. New York, NY: Routledge.

Denzin, N. K., & Lincoln, Y. S. (Ed.) (2000). *Handbook of qualitative research* (2nd ed.). Thousand Oaks, CA: Sage.

Elo, S., & Kyngäs, H. (2008). The qualitative content analysis process. *Journal of Advanced Nursing, 62*(1), 107–115.

Freire, P. (1972). *Pedagogy of the oppressed*. Harmondsworth: Penguin.

Gergen, K. J. (2014). From mirroring to world-making: Research as future forming. *Journal for the Theory of Social Behaviour, 45*(3), 287–310.

Haraway, D. (1997). *Modest_witness@second_millennium.FemaleMan_meets_OncoMouse: Feminism and technoscience*. London: Routledge.

Palmer, P. J. (2017). *The courage to teach: Exploring the inner landscape of a teacher's life* (20th anniversary ed.). Hoboken, NJ: Jossey-Bass.

Polanyi, M. (2009). *The tacit dimension*. Chicago, IL: University of Chicago Press.

Van de Putte, I., De Schauwer, E., Van Hove., G., & Davies, B. (2020). Violent life in an inclusive classroom: Come on, read, Peter! Moving from moral judgment to an onto-epistemological ethics. *Qualitative Inquiry, 26*(1), 60–70.

On Relational Ontology and the Good Life

To bring the outcomes of inquiry with co-researchers as presented in Chap. 3 into a generative conversation with state-of-the-art complexity thinking, in the next two chapters I perform a reading of emerging axiological insights (this chapter) and praxeological insights (Chap. 5). In doing so, I continue the diffractive methodology introduced in Chap. 3 – understood as reading insights through one another (Barad 2007; Bozalek and Zembylas 2017) – yet in a slightly different mode. Whereas in Chap. 3, narratives of co-researchers were read through each other, I now read the data this process generated through multiple theoretical insights (Ceder 2019; Lenz Taguchi 2012; Mazzei 2014). Chapters 4 and 5, therefore, engage extensively with theoretical literature, and they do so through what Ceder (2019) calls a pragmatic approach: the focus is not on distinct philosophers but on 'the creative entanglements that could be used to solve the research problems at hand' (p. 49). I understand my task in these chapters, thus, not as to distill and present some pure form of the work of certain scholars, but as 'to engage aspects of each in dynamic relationality to the other' (Barad 2007, p. 93) so that 'the relationality of them can contribute to solving the philosophical problem' (Ceder, p.54).

The first step I take in this chapter is to consider how entanglement can be experienced in various forms, some more positive than others. For this, I engage especially with the work of Ian Hodder, who explores how entanglements have both enabling and entrapping effects, and the concepts of resonance and alienation as introduced by Harmut Rosa. Building on this understanding, I subsequently perform a diffractive reading of the axiological insights articulated by co-researchers. The scholars whose ideas I will primarily engage with in this process are Alasdair MacIntyre, Paulo Freire, Gert Biesta, Karen Barad, Simon Ceder, and Daniel Wahl. The first three scholars in this list help to build forth on Rosa to argue for a pedagogical orientation toward flourishing as an entangled phenomenon (i.e. rather than an individual achievement at the cost of another's loss) and I will refer to this as entanglement-orientedness. Barad, Ceder, and Wahl, in turn, help to read into the subtleties of this idea through the lens of a relational ontology (e.g. by proposing post-anthropocentrism rather than anthropocentrism, by emphasizing that entanglement-orientedness can both be experienced as a rational imperative and a deeply intuitive incentive, and by exploring how entanglement-orientedness can manifest both in awareness and action). Some of these scholars have played an important role throughout the earlier chapters of my inquiry, yet Hodder, MacIntyre, Rosa, and Ceder enter the stage for the first time here. Hodder is an entanglement theorist who attracted me for, unlike Ingold and Barad, he emphasizes that entanglement has a dark side, which, so I have found, is essential in

the process of articulating pedagogical intentions. Ceder's work attracted me as he, being a
follower of Barad, manages to provide a thorough account of what the insight of entanglement
implies for our understanding of educational
relations, and as such is an excellent recent example of thinking about education through the lens
of relational ontology. I thank my engagement
with the moral theorizing of MacIntyre and Rosa
to lively discussions with acquainted researchers,
both of which were incremental in shifting focus
from an axiology focused on individual students
to one focused on relational quality.

Entrapment and Resonance

Having established the understanding that students are entangled in complex societal challenges, I would like to start this reading by
suggesting that this entangledness is not itself a
positive or negative phenomenon. We might see
it, rather, as an experience that can manifest itself
in myriad ways and which can have both constructive and destructive effects on ourselves and
others. In this light, it is interesting that co-
researcher group 2 articulated, in diffractive step
1.2, that it is important to question if students'
particular entanglements are good for them and if
it is good for students to be confronted with particular elements of their teacher's entangledness.
Similarly, group 3 articulated, in diffractive step
2.2, the aim to strive toward "light/positive"
rather than "dark/negative" entanglements. In my
experience, the theorizing of Ian Hodder (2014)
can help to create some clarity at this point. In his
work, Hodder focuses on the entanglement
between humans and the material things humans
produce. To understand how we are entangled
with the things we produce, Hodder distinguishes
two forms of dependence. The first form is that
our entanglement with things is enabling, as in
using a car enables us to move across great dis-
tances in a short amount of time. Seen this way, a
car is something we positively rely on and the
history of technology a huge success story; we
have become able to visit loved ones all around
the world and to meet them online, and we live in

houses with such comforts as central heating,
refrigerators, baths, and so forth. The second
form, however, is that the entanglement with
things is constraining, as in using a car forces me
to earn and spend money, makes me complicit in
air pollution, and if I am driving to and fro a
party, I am not permitted to drink alcohol. In this
sense, the history of technology is a story of
increased entrapment; we depend on more and
more material things that in turn depend on our
investment of time, attention, money, and energy,
and these resources are, in fact, very valuable to
us and often scarce. Hodder's analysis evokes the
question, therefore, if it is truly worthwhile to
strive for material comfort and technological
innovation as much as we do. This question
becomes especially pressing if we realize that the
things we produce to enable us are always imperfect; the car gets broken and every year more
advanced versions enter the market. This instabil-
ity and ongoing innovation of material things
have resulted in worldwide networks of produc-
tion, trade, destruction, and recycling. It is not
hard to see, and so Hodder emphasizes with care,
that next to all the gains this too has a dark side:
our modern lifestyle depends on inhumane work
conditions in third-world countries, unfair distri-
bution of wealth, and tremendous damage to the
ecological systems of the Earth.

Our entanglements, thus, have both enabling
and constraining effects, both on ourselves and
others. If the constraining effects overshadow the
enabling effects, and if we cannot simply step
away from this, we may start experiencing an
entanglement as an entrapment. Let us consider
the diffractive script 'The sustainable school' as
an exemplary case (n.b. I choose this particular
diffractive script here as it invites consideration
of both our entanglement with humans and with
other forms of life). The two students who want
to start a sustainability committee are unhappy
about the effects our lifestyle has on nature, yet
cannot simply step away from this, for they
depend on those very things and systems that
harm nature. The school they attend, the house
they live in, the food they eat, the computers,
lamps, and heating systems they use, all play
their part in the unsustainability-entrapment.

Although these students want sustainability, they are in many ways trapped in patterns of unsustainability, and even if they manage to drastically reduce their ecological footprint the ecosystems of our world are far from saved. Being entrapped as such is not at all an easy position to be in. In fact, depression, socio-ethical paralysis, and loss of well-being triggered by awareness of and complicity in the ecological crisis – eco-anxiety in short (Pikhala 2018) – has become a serious contemporary problem. The way we are entangled, and the degree to which we experience our entanglements as enabling and constraining, as freeing and entrapping, thus affect us psychologically and behaviorally. This brings me to the theorizing of Rosa (2017), and his concepts of alienation and resonance.

For Rosa (2017), similarly to Hodder, there is a dark side to the success story of the wealth and technological advancement of modern societies. Entangled as we are in making the world more available, accessible, and attainable through economic growth, technological innovation, and self-improvement – so he observes – we tend to find ourselves constantly under pressure and short of time. We moderns resemble, according to Rosa, 'a painter who is forever concerned about improving his materials – the colours and brushes, the air condition and lighting, the canvas and easel, etc. – but never really starts to paint' (p. 443). Note how this analysis resembles the drive behind the initiation of The Bildung Academy as described in Chap. 1 and the educational critique developed in Chap. 2, which can very well be summarized by stating that we tend to be preoccupied with equipping students for predicted future lives whilst forgetting the lives they live today as well as the possibility of alternative futures. More important for my current concerns, however, is how an entrapping entanglement manifests itself. Rosa argues, on this matter, that in our preoccupation with our equipment with resources we run the risk of alienation: 'a particular mode of relating to the world of things, of people and of one's self in which there is no responsivity, i.e. no meaningful inner connection' (p. 449). It is not, so he argues, that we lose our relationality (i.e. we are still entangled)

but our relations tend to become marked by the absence of 'a true, vibrant exchange' (p. 499). This absence, Rosa elaborates, can show itself in two contrasting ways. We either move toward indifference (see, also, Dohmen 2007) or hostility. We either lose our inspiration, become obedient followers, become the oppressed (Freire 1972), perhaps even grow into burnout or depression, or we become frustrated aggressors, perhaps become oppressors ourselves, in violent opposition with the world. Unfortunately, we live in a time in which an increasing number of young people, students in our schools and universities, experience some form of depression or burn-out (Walburg 2014). Unfortunately, also, the COVID-19 pandemic provides an excellent example of how entrapment – for that is, indeed, how many of us experienced the periods of lockdown – can lead to mental health problems among students (see, amongst a rising number of recent studies, Son et al. 2020) and, on many occasions throughout the world, to receptiveness for radical complot theories and outbursts of violent protest.

Yet, alienation, either in the form of indifference or hostility, is not the only possible response to the experience of entrapment, and neither is it the response observable in the exemplary story of the initiation of The Bildung Academy or the diffractive script 'The sustainable school'. Rather than a story of indifference or aggression, the story of The Bildung Academy is, first and foremost, a story of young people coming together for a constructive process of creating new ways of educating. And rather than becoming indifferent or aggressive, the two students in the diffractive script 'The sustainable school' take the constructive step of collaboratively initiating a sustainability committee. Entrapment can, thus, also give rise to a very different response and experience, one in which we do not become indifferent, or position ourselves in violent opposition, but in which we engage, create, and learn from a position of hope. Rosa refers to this other kind of experience as resonance, which he describes as a dual movement of a←fection, of something other that touches you (i.e. being shaped), and e→motion, your response that touches something other (i.e. being a shaper). Resonance, in other words, is Rosa's conception of an experience in which we embrace our entangledness, in which we are simultaneously open to the otherness we are entangled with and speak with our own voices, and through this actively commit to the ongoing formation of the world. Such a commitment, notably, is not one to abolishing difference but to collaborative inquiry into difference, for without difference, so Rosa emphasizes with care, there can be no experience of resonance, only sameness everywhere and thus no creative vibration.

Entanglement comes, thus, in many forms. Without claiming to be exhaustive, I can conclude that entanglements have both enabling and entrapping effects. Entanglements can both lock-in and set-free. What hangs in the balance, is how we respond to these effects, and that, so shall be the central line of thought throughout this chapter, is where pedagogy comes into play.

Entanglement-Orientedness

Reading the axiological insights articulated by co-researchers for the three diffractive scripts, one pattern of insight drew my attention first, namely: they all emphasize the existential task of living together. When a student shouts out 'I bet those f*cking Moroccans did it!', Pien, Ronald, and Sandra aim for the openness to share feelings, thoughts, and believes with each other, the safety of doing so in a way that does not exclude anyone from existing and living together, and the experience of dealing with the challenges and differences we encounter constructively. When two students want to start up a sustainability committee, Jacob, Astrid, Jens, and Kasper aim for respect and openness for diverse points of view concerning sustainability, yet without losing the capacity to critically question their grounds (e.g. as being unscientific), and for collaboratively transcending "bubbles of sameness" to grow into more realistic and sustainable perspectives together. And, when an American student in a boarding school in the Netherlands responds to a mocking cartoon of Donald Trump

by defending Trump's politics and criticizing Dutch media, Aafke, Steven, Irene, and Elmarie see this as an opportunity to practice listening to each other, to respect and value other points of view, to express oneself, and to critically question oneself and each other. Each group in its own way emphasizes the importance of being open and caring toward otherness, and the importance of thinking for oneself and speaking one's own voice. Together, perhaps, these two ways of being – receptive and responsive, caring and rational, open and critical – provide a helpful pedagogical orientation. In the paragraphs to come, I perform a diffractive reading of this idea.

To start, let us return to Rosa (2017), whose concept of resonance so clearly integrates these two ways of being (i.e. receptive and responsive) into one experience. Interestingly, for my purposes here, Rosa's theorizing on alienation and resonance departs from a fundamental critique of ethical pluralism, which he considers to be the basic cultural condition of modernity (p. 442):

> whether one should abide by a religious belief, and if so, by which one, whether one should strive to develop political, artistic or intellectual capacities, whether one should marry and have kids or not, and all the other small and big questions about what kind of life one should lead, about leading a life as such – e.g. whether music should be important, whether literature should be a part of life, whether the town or the country is preferable, whether the local soccer team was important or not – were turned into strictly private questions. *You'll have to find out for yourself!* is the standard answer to all of them.

It is, of course, in many ways a victory that more and more people have increasing freedom to decide for themselves how they want to live (e.g. what to study, where to live, what work to do, who to marry), yet, this very freedom also underlies, in Rosa's analysis, the modern human's resemblance to a painter who never really starts to paint. Lacking an answer to what the good life consists of, the overruling rational imperative of modernity has become, so he argues (p. 443): 'secure the resources you might need for living your dream (whatever that might be)!'. It is not that Rosa is essentially against increasing the scope of our money, wealth, options, or capabilities, yet he emphasizes that if these become ends in themselves rather than means for 'a resonant mode of being' (p. 453), we run the risk of alienation. For Rosa, therefore, the good life is not a matter of individual achievement, nor a private matter, but in its essence a matter of relational quality. This suggestion parallels the move from an ontology of individualism to one of relationality: if we are entangled, and if our individuality emerges in relationality, then so is the quality of our lives a mutual agenda and accomplishment! Rather than primarily focusing on helping students to become the best versions of themselves, the orientation of pedagogy of entanglement could be to help students co-create and nurture nourishing and stimulating relationships.

These considerations bring me to the work of perhaps one of the most influential moral philosophers of our time: Alisdair MacIntyre. Like Rosa, an important starting point for MacIntyre is to be critical of the idea that the good life is a private matter (see, especially, 1990, 2007). His criticism is directed at 'those who see in the social world nothing but a meeting place for individual wills, each with its own set of attitudes and preferences and who understand that world solely as an arena for the achievement of their own satisfaction' (2007, p. 25). This type of morality is, for MacIntyre, a serious symptom of the moral disorder of modernity, for it renders it impossible to have adequate public discourse on areas of moral concern that affect us all, such as the allocation of resources to the poor and rich or to the young and old. The majority of MacIntyre's work is devoted to exploring what virtues might help us to overcome our modern predicament. Browsing his bibliography, one of his later works – *Dependent rational animals: Why human beings need the virtues* (1999) – attracted me especially, for its title hints toward the acknowledgment of a relational ontology. This book was published 18 years after MacIntyre's foundational work *After virtue* (2007, first edition published in 1981) and his work had up to then primarily focused on what he calls the virtues of independent practical reasoning: the willingness and ability of an individual human being to eval-

uate and (re)direct reasons and desires, imagine and choose between possible futures, and make oneself accountable for one's endorsements. In *Dependent rational animals*, however, MacIntyre's plea for the independent practical reasoner is enriched with a thorough consideration of our fundamental dependencies. It is at this point in his work, I believe, that a relational ontology starts being recognizable. The facts that we, as human beings, are vulnerable to many kinds of affliction and throughout our lives in many ways depend in our survival and flourishing on others, so he argues, 'are so evidently of singular importance that it might seem that no account of the human condition whose authors hoped to achieve credibility could avoid giving them a central place'. Acknowledging our dependency, acknowledging, in other words, that our flourishing is not only of our own making, moves MacIntyre to articulate the virtues of acknowledged dependency as 'the necessary counterpart to the virtues of independence' (p. 120). The fact that we depend, throughout our lives, on our parents, friends, teachers, neighbors, physicians, and so forth, calls upon us, so he argues, to in turn care for others that are presented to us in need. We cannot become nor stay independent practical reasoners without the care of others, and a true independent practical reasoner would, therefore, naturally seek to contribute to those very relationships that we all depend on. The virtuous human being, for MacIntyre, is thus both rational and caring, both receptive to the needs of others and accountable for one's endorsements, both speaking with one's own voice and listening to others. The implications for education are, according to MacIntyre, profound, as he concludes (p. 160):

> We do indeed as infants, as children, and even as adolescents, experience sharp conflicts between egoistic and altruistic impulses and desires. But the task of education is to transform and integrate those into an inclination towards both the common good and our individual goods, so that we become neither self-rather-than-other-regarding nor other-rather-than-self-regarding

MacIntyre's theorizing thus helps to argue for a pedagogical agenda of integrating self-orientedness and other-orientedness into what I suggest to call *entanglement-orientedness*. In short, the argument goes as follows: as my good and that of the other are not separate and co-dependent (i.e. they are entangled), it would be wise to try to nurture relationships from which both can emerge. This argument bears resemblance to the recent work of Gert Biesta (2017, 2020), whose theorizing has been an important inspiration throughout my inquiry. Education, as Biesta argues, always impacts, either intentionally or unintentionally, students' subject-ness, that is, it impacts 'the individual who acts (or decides not to act)' (2020, p. 99). The majority of Biesta's work is devoted to a consideration of what our pedagogical aims are to be in the domain of subjectification, and the conclusion he arrives at resembles the perspective of entanglement-orientedness. Education as subjectification, according to Biesta, is to be oriented toward the never-resolved existential challenge of 'trying to stay in the difficult "middle ground" in between world-destruction and self-destruction' (2020, p. 97). Introducing these two opposing, destructive modes of being, he explains (2020, p. 96–97):

> From the perspective of our intentions and initiatives, the encounter with resistance generates a degree of frustration. Out of such frustration we could try to push harder in order to overcome the resistance we encounter. This is sometimes important for our initiatives to arrive in the world, but there is always the danger that if we push too hard, we may destroy the very world in which we seek to arrive. If, then, at one end of the spectrum we encounter the risk of world-destruction, at the other end we find the existential risk of self-destruction: when confronted with this double-bind, out of frustration, we step back and withdraw ourselves from the situation.

Entanglement-orientedness entails, thus understood, 'a way of being together that seeks to do justice to all partners involved' (Biesta 2017, p. 14). The idea is to try to move beyond the realm of contest or competition, in which my win is the logical consequence of your loss and I either care for the other or myself, and toward the realm of cooperation, in which we aim to win together. Entanglement-orientedness – so is the suggestive conclusion of this part of my read-

ing – offers somewhat of a meta-narrative in the dimension of axiology, which parallels the move from an ontology of individualism to a relational ontology. Let us look, therefore, a bit closer at this perspective through the lens of relational ontology. There are at least four important points to make.

As discussed in Chap. 2, the premise of entanglement gives rise to the relational ontological interpretation that the experiences of self and other emerge in relationality rather than exist prior to it (Barad 2007; Ceder 2019). Following this logic, and so is the first point, I consider it important not to understand entanglement-orientedness as an interest in some sort of compromise between the desires of separate, fixed individuals or groups, but rather as a strive toward the entangled flourishing of entangled lives. To speak of a "middle ground", as in the choice of words of Biesta, is, in fact, a bit confusing in this sense, for it invites the interpretation that this is a place we go to – as individuals previously existing elsewhere – so to enter into a relationship. Rather, I understand relationality to be where we exist in the first place, and understood as such the "middle ground" refers, in fact, to relationality of a certain dialogical quality from which individual and collective well-being and creative potential can emerge simultaneously (e.g. receptive and responsive, open and critical, caring and rational). The extremes of self- or world-destruction are, then, likewise not experiences outside of relationality (n.b. Biesta's usage of the words "withdrawal" and "arrival" do suggest so), but rather typologies of particular kinds of relationality that obstruct rather than enable the experience of individual and collective well-being and creativity (e.g. the experiences of indifference and hostility). To be more explicit: both a culture of fundamentally suppressing communal order and responsibility for the sake of individual freedom and one of fundamentally suppressing individuality for the sake of a totalitarian regime, can be considered to work against the emergence of entangled flourishing. What we rather need, then, is 'a collective sense of us […] in which individualities are engendered' (Elwick 2020, p. 149). For are not the experience of community in a way that invites self-expression and the expression of individuality in a way that strengthens communal relationships the true signs of the emergence of entangled flourishing?

If entanglement-orientedness, then, is not about compromising between fixed positions, it inevitably necessitates, sometimes, processes of transformation (Wahl 2016; Rosa 2017). Consider, for instance, Hodder's argument that we have become more and more entrapped in our dependency on the material things we produce, such as cars. Two key entrapping dynamics in this example are: (1) owning a car is expensive and maintenance demanding, and (2) owning a car increases one's negative impact on nature. Two key developments that work toward a way out of this entrapment are: (1) the emergence of car-sharing initiatives, which takes away the necessity to purchase and maintain a car yet provides an affordable alternative, and (2) the transition from gas toward more sustainable energy sources. These developments are not just compromises between the car owner, his wallet, and the climate, they are profound transformations of their relationality. This includes, for instance, the transformation of the desire to own a car to the desire to use a car when needed, and the transformation of the desire for mobility to the desire for sustainable mobility. Another way to make this point is to look at it from the perspective of Freire's pedagogy of oppression (1972). Relationships of oppression are, in their very nature, win-lose relationships; the power of one is constituted by the powerlessness of another. The way out, I agree with Freire, is not some kind of compromise between the oppressed and the oppressor, but a profound transformation of their relationality into another relational pattern. Entanglement-orientedness, then, is about opening ourselves to the possibilities of life-in-becoming and about the ambitious aim to co-create a world in which we can all win together.

Secondly, whereas scholars such as MacIntyre and Freire still focus on anthropocentric relationality – i.e. human-human entanglements – frontiers of relational ontology and relational pedagogy, such as Barad, Ceder, Rosa, and Wahl, explore entanglement through a post-anthropocentric lens (i.e. entanglement in a more-than-human-world). The problem with anthropocentrism, as Ceder formulates it, is that 'the human/nonhuman distinction creates the possibility of objectifying and the thing that is objectified is not seen as an ethical subject' (2019, p. 37). Reading MacIntyre, for instance, I was impressed by his consideration of the degrees to which we depend on other human beings, and the moral implications this entails, yet I also felt as if his arguments were incomplete as he fails to do the same for our dependency on the ecosystems we are part of. Reading Freire, similarly, I was touched by the energy he devotes to the liberation of oppressed human beings, but also struggled with the question: what about all the other forms of life that live, to a large extent, under a human regime? Reading Rosa, in contrast, did not trigger such feelings, for from his descriptions it is abundantly clear that resonance is a relational quality that can just as well be experienced with a fellow human being as, for instance, with an ocean, a piece of wood, or a dog (and I shall return to the dog momentarily).

I consider including the more-than-human into our considerations to be immensely important, for we are, indeed, entangled with, and thus dependent on, not only other human beings but also the microbes in our stomachs, the air-regulating trees in the forest, the biodiversity in the ocean, and so forth. Furthermore, many of the contemporary challenges we face, such as global warming and virus-pandemics, are deeply related to how we, as modern human beings, relate to the more-than-human world. As Barad puts it strikingly: 'a humanist ethics won't suffice when the "face" of the other that is "looking" back at me is all eyes, or has no eyes, or is otherwise unrecognizable in human terms' (2007, p. 392). We have to consider, in Ceder's words, nonhuman life as ethical subjects, subjects we depend on and who depend on us. If we reread MacIntyre's reasoning through this insight (see, also, Hannis 2015), nonhuman life deserves our care just as does our fellow human being. Such a plea for including the more-than-human into our conception of the common good is, indeed, gaining traction, and is, for instance, clearly observable in the fields of holistic education (e.g. Hart 2014) and education for sustainable development (e.g. Lotz-Sisitka 2017). Entanglement-orientedness, I believe, ought to be understood not only as an orientation toward co-constituted human flourishing but also as the ambition to let our entanglement with the more-than-human be a win-win story rather than a win-lose story.

Thirdly, although the line of reasoning presented above can appear very rational and calculated – as we depend on each other we should care for each other – I care to emphasize that entanglement-orientedness can also be experienced as a strong intuitive incentive. For, do we not all know from personal experience that much of our actions are not calculated as such? If we do something for another, for instance if I take my dog for a walk, it is not merely a rational decision (e.g. 'if I do not walk with him now he might pee in the living room') but also an intuitive desire for him to relieve the pressure in his bladder. It is, what Freire (1972) would refer to not just as an act of reason, but an act of love. As he argued passionately (1972, p. 62–63): 'no matter where the oppressed are found, the act of love is commitment to their cause – the cause of liberation. […] If I do not love the world – if I do not love life – if I do not love people – I cannot enter into dialogue'. Intuitively, I do not want my dog to suffer, and this is not merely the result of a rational calculation of the mutual dependencies between us. Simply: it hurts me to see him suffer and it feels good to see him happy. Likewise, I intuitively do not want to suffer myself; I prefer to be healthy rather than sick, with loved ones rather than lonely, devoted to a meaningful cause rather than without direction, and so forth.

For me, the intuitive desire for the well-being of all life resonates with Buddhistic traditions (e.g. Dalai Lama 1999; Ricard 2015), but I am aware that similar notions can be found in many other spiritual and philosophical traditions. For this chapter, it is perhaps most fruitful to highlight how the intuitive desire for well-being also plays an important role in the theorizing of Barad, who articulates it poetically when she writes: 'a delicate tissue of ethicality runs through the mar-row of being' (2007, p. 396). For Barad, as we have seen, to live is to be simultaneously a shaper of and shaped by the world and responsibility, therefore, an existential experience that is unavoidable (see, on this point, also Biesta 2013; Ceder 2019). We may choose how to respond, but we do not choose responsibility itself. When faced, for instance, with the suffering of my dog, I *am* responsible, I cannot not respond. I can decide to ignore my dog or to go away not to be bothered, but these are still responses with consequences. In other words: "doing nothing" is an action that re-establishes ongoing dynamics (e.g. the increasing suffering of my dog, at some point inevitably leading to an accident on his behalf). Ethics, in this line of reasoning, is about 'responsibility and accountability for the lively relationalities of becoming of which we are a part' (Barad 2007, p. 393). All our actions, whether we want it or not, co-constitute the world we are part of, and consequently, in Barad's words, the intuitive desire to respond to suffering and protect well-being 'runs through the marrow of being'. Entanglement-orientedness can be seen, therefore, not only as a rational, but also as a deeply intuitive ideal, driven by the desire for joy and peace rather than war and suffering, not just for me or you, but as entangled phenomena.

What the ideas explored throughout this chapter share, is that they redirect the focus from me-against-you to the relationality in which we co-exist, and consequently propose to strive toward a 'responsible entangling within the world' (Edwards 2012, p. 533). I believe this might be the main concern of pedagogy of entanglement. There is, then, a fourth and last point to make. Following this logic, namely, entanglement-orientedness is not to be understood as something a student can acquire and possess, but as what Biesta refers to as 'a never-resolved existential challenge' (2020, p. 97), that has to be practiced over and over again. Furthermore, if entanglement-orientedness is a practice, in this line of reasoning it is to be understood as a communal one. According to relational ontology, there is no separate entity that possesses the orientation toward entangled flourishing and which brings it to an encounter. Rather, there is relationality in which it is accomplished and experienced (Barad 2007; Ceder 2019). The perspective of entanglement-orientedness resists, then, educational tendencies to only/primarily focus on the individual growth of students as measured by a growing list of competencies (e.g. the trend of twenty-first-century skills that was briefly discussed in Chap. 1). A student can, of course, experience and demonstrate commitment to the cause of entangled flourishing, but this is never simply that singular student's accomplishment nor one that is established once and for all. Am I to further develop a pedagogy of entanglement, I need, thus, to build an understanding of what it means to invite and practice entanglement-orientedness collaboratively. I take on this task in the next chapter when I shift focus to the dimension of praxeology. Before I go there, however, let us reconnect to the axiological insights articulated by co-researchers once more, for, so shall be my argument, these help to translate the perspective of entanglement-orientedness into a more practical axiological orientation for teaching students in the face of complex societal challenges. Co-researchers' aims are articulated, namely, within the interrelated dimension of awareness and action, which, so shall be my suggestion, together can be a manifestation of entanglement-orientedness.

Entanglement-Awareness and Hopeful Action

Concerning awareness, in the diffractive script 'The multicultural classroom', my co-researchers hope for increased awareness of the origins and effects of stereotyping statements, and of the perspective that we all belong in this world together and therefore need shared rules, norms, values, and responsibilities. Similarly, in the diffractive script 'The sustainable school', my co-researchers strive for increased awareness of the importance and complexity of sustainability, of societal dynamics related to sustainability (e.g. trends and bubbles and the matter of facts versus fake news), and of our own thoughts and actions concerning sustainability. In the diffractive script 'Mock and

prejudice', lastly, my co-researchers aim for increased awareness of own and others' values, how we respond if values are threatened, how values are co-shaped by media-framing and environment, if perhaps we hold identifications that we would rather let go of, and how our actions impact others. All these exemplary aims are concerned with an increased awareness of (1) what we are entangled in/with, (2) in what particular way we are entangled and along what path this has become so, (3) how others are entangled and what their stories of becoming are, and (4) our impact on, dependency on, and responsibilities toward others. For me and my co-researchers, therefore, a useful way to summarize these aims, is as an aim toward *entanglement-awareness*. To use Ceder's words (2019, p. 98): 'each subject needs to discover the human and nonhuman relations s/he is part of'.

It is important, at this point, to bring into remembrance the critique on representationalism as presented in Chap. 3. In light hereof, entanglement-awareness is not achieved passively, by representing from the outside, but actively by intra-acting within (Barad 2007; Polanyi 2009). We are to acknowledge, therefore, that our awareness is always partial and colored by our historicity. We can, however, become less and less ignorant of the myriad and dynamic particularities of our entangledness in the world. I agree wholeheartedly with Freire when he writes (1972, p. 63): 'at the point of encounter there are neither utter ignoramuses nor perfect sages; there are only people who are attempting, together, to learn more than they now know'. Another way of framing all this is to argue that aiming for entanglement-awareness is not a matter of reaching "pure" or "absolute" awareness, but an ongoing quest of making sense of this dynamic, and indeed sometimes rather confusing reality we find ourselves in. What, then, does this sense-making entail? I move on to suggest that, at least, this has to do with connecting the personal to the systemic, the past to the future, and criticism to understanding.

Let me start with connecting the personal to the systemic. On the one hand, we can observe that entanglement-awareness enables us to 'lead storied lives' (Connelly and Clandinin 1990); we become aware of the particular events that make our lives uniquely ours, and the particular choices and commitments that provide us with a sense of identity, a sense of being someone in the world (Kroger and Marcia 2011). On the other hand, we can consider how our lives are part of 'a mesh-work of interwoven lines of growth and move-ment' (Ingold 2010, p. 3); the systems and bureaucracies we depend on, the rules, norms, and values that mediate our participation in the world, the patterns of communication between individuals and between groups, the plurality of perspectives and concerns people hold on to, hierarchies of power, dynamics of entrapment, and so forth. Entanglement-awareness, as I see it, integrates such systemic awareness (Wahl 2016) with a lively sense of one's path through the world (Ingold 2008). If we are very much aware of our own paths, yet unaware of the paths of oth-ers, and of the systems we are immersed in, we have likely been self-rather-than-other-oriented, and in the reverse scenario it is the other, and the bigger picture, that has likely consumed the majority of our attention. Yet do we demonstrate an integrated awareness of our own path and its interweaving in the wider world, we can interpret this as a sign that we are practicing entanglement-orientedness.

Entanglement-awareness, also, can be seen as insight into how past and future are connected. The perspective of entanglement highlights that life is relationally produced (Barad 2007), and therefore focuses attention on the here-and-now, on the tangible reality of the interweaving of lines of becoming (Ingold 2010). Yet this inter-weaving is an ongoing phenomenon, not some-thing that just starts out of nothing and goes nowhere, but a movement with a history and a future. Striving for entanglement-awareness entails, thus seen, two simultaneous efforts: the effort to understand how we have arrived at where we are today (i.e. the events and processes through which our entangledness in the world took form) and where we might be going hence-forth (i.e. the possible futures that we might col-laboratively create). We can, if we pay close attention to experiments around us (e.g. ecologi-cal housing projects), even recognize what Wahl calls 'pockets of the future in the present' (2016, p. 54) that provide a glimpse of what the future may look like if we choose to embark on a certain path.

If we understand entanglement-awareness thus as connecting the personal to the systemic and past to future, it opens the door, also, to be both critical and understanding. Critical, in a Freirean sense (1972), in the sense of becoming aware of particular entrapments in our mutual lives and the destructive or oppressive forces at work in them (e.g. the sustainability-entrapment). S/he who strives for entangled flourishing shall feel naturally inclined to resist self- and world-destruction and to advocate forms of entangle-ment in which the different parties involved co-constitute each other's well-being and creative potential. Understanding, secondly, in the sense of recognizing that although there are destructive forces at work, these are the result of a relation-ally produced history and thus there for a reason and not simply one person's fault. Of course this does not mean that we should not be critical, but being critical without trying to understand can itself be considered an aggressive approach, which rather than resonance and transformation invites alienation and hostility. An important question always is, then, what one does with one's awareness. It matters, greatly, if awareness manifests as indifference or hostility, or if we seek dialogue and progressive self- and world transformation. This brings us to the other dimen-sion in which co-researchers articulated inten-tions: actions.

Concerning students' actions, the three groups of co-researchers respectively hope that their stu-dents (1) make their own, humane choices, take responsibility for these, and deal with conflict/differences constructively, (2) express their opin-ions qualitatively (e.g. with sound arguments), transform their interests and enthusiasm into action, and grow toward sustainability collabora-tively, and (3) shape their entanglement with each other in a light/positive way (e.g. respectfully lis-tening to each other and expressing themselves). The type of actions that co-researchers value are, so we can now interpret, an expression of entanglement-orientedness; they embody the effort to conscientiously speak one's own voice in co-creative dialogue with other voices. Thus avoiding both self- and other-neglect, this mode of being and acting in the world can be considered to be truly hopeful. Hopeful, in the sense of being committed to a better-shared world (Wahl 2016). According to Freire (1995, see, also, Dasberg 1983), hope is an ontological need that drives education. It is only for we believe in and strive toward a better-shared world that pedagogy of entanglement makes sense. Yet, as Freire stressed very well (p. 2), 'just to hope is to hope in vain'. Hope ought to be manifested in action. What kind of actions, then, can be considered hopeful expressions of entanglement-orientedness?

I would like to suggest that we can consider at least three kinds of actions that are hopeful expressions of entanglement-orientedness, namely: acts of conservation, acts of adaptation, and acts of regeneration. The difference is perhaps most clearly explained along the lines of an example. Let us reconnect, therefore, once again to the diffractive script 'The sustainable school', in which two students initiate a sustainability committee. Concerning acts of conservation, first of all, it is worth noting that critical pedagogical accounts, such as those of Freire and Wahl, often tend to focus on change and transformation of the undesired into the desirable, rather than on conserving and protecting the desirable that is already around us. In light hereof, environmental education researcher Chet Bowers (2003), provides a welcome plea for what he calls mindful conservatism; we ought, so he argues, to ask ourselves not only the question of what we want to change but also what we want to conserve, and then take responsibility for this very conservation. From the perspective of entanglement-orientedness, acts of conservation strengthen and nurture such relationships from which mutual flourishing emerges. For instance, the sustainability committee in the diffractive script 'The sustainable school' might decide to join forces with the biology teachers of the school to investigate what diverse types of desirable vegetation grow around the school parameters (desirable,

for instance, for they absorb CO_2, attract insects, contribute to bio-diversity…), to ensure they have enough space and nutrients to flourish. If we truly want to work toward entangled flourishing, we do however need more than conservation of the good that is already there. We need, also, to move away from undesirable relational dynamics toward desirable ones. This is where acts of adaptation and regeneration come into play (Rosa 2017; Wahl 2016). From the perspective of entanglement-orientedness, acts of adaption tweak existing relational dynamics to reach a better harmony of needs involved. For instance, the sustainability committee might convince the school management to improve the isolation quality of the windows of the school building, so that less heating is needed to keep students and teachers warm and comfortable. This is an important step toward a more sustainable school building, yet it does not resolve the entrapment of being dependent on unsustainable energy sources. The sustainability committee can also, therefore, start collecting money to invest in solar panels on the roof to embark on the path toward more sustainable energy sources. Such a move would be an act of regeneration, for this replaces a relational pattern of unsustainability with a more sustainable one. From the perspective of entanglement-orientedness, acts of regeneration thus move toward transforming win-lose-relationalities into win-win-relationalities.

Students' actions can be hopeful, then, in myriad ways, depending on the specific situation. If we look back at the three diffractive scripts, we can observe multiple destructive risks. The risk of global warming. The risk of increased polarization and conflict between Dutch citizens with and without a Moroccan background, or between Trump supporters and Trump adversaries. The more immediate risk, also, of indifference or hostility for the sustainability-concerned students, the student who makes the Moroccan-statement and the Moroccan students who are present, and the Trump-supporting student. If we manage, as teachers, to elicit actions in our students that work against these risks and give rise, instead, to an experience of resonance (Rosa 2017) or liberation (Freire 1972), we truly deserve to be regarded as agents of hope. Students' action, for instance, of initiating a sustainability committee. The action, also, of the Trump-supporting student to make him-/herself heard despite being a minority in the classroom. Any action, in other words, that brings to the foreground that which is entrapped – be it self, other, or planet – and promotes doing right by all partners involved. Such actions bear witness of the insight, in line with Rosa (2017) and Wahl (2016), that flourishing does not entail endless possibility and growth, but rather is about mutual well-being and creative opportunity within community and its boundaries. I truly believe that the more we become collectively aware hereof, and learn to live and act accordingly, the bigger our chance of leaving a world behind for future generations that we can be proud of.

In closing, the confrontation with entrapment can just as well make us feel guilty, hopeless, and lost, as it can spark feelings of responsibility, connectedness, and purpose, and our pedagogy can be the difference. The difficult news of the premise of entanglement is that we are complicit in patterns of entrapment, but the exciting news is that together we can act toward a world that is shared better than it is today. Pedagogy of entanglement can seek to generate the experience that we can make a difference that matters, that our actions can conserve that which is valuable, adapt that which is off-balance, and regenerate that which is destructive. Such a pedagogy would turn education into a true manifestation of hope.

References

Barad, K. (2007). *Meeting the universe halfway: Quantum physics and the entanglement of matter and meaning.* Durham: Duke University Press.

Biesta, G. J. J. (2013). *The beautiful risk of education.* Boulder, CO: Paradigm Publishers.

Biesta, G. J. J. (2017). *The rediscovery of teaching.* New York, NY: Routledge.

Biesta, G. (2020). Risking ourselves in education: Qualification, socialization, and subjectification revisited. *Educational Theory,* 70(1), 89–104.

Bowers, C. A. (2003). *Mindful conservatism: Rethinking the ideological and educational basis of an ecologi-cally sustainable future.* New York, NY: Bowman and Littlefield Publishers.

Bozalek, V, & Zembylas, M. (2017). Diffraction or reflection? Sketching the contours of two methodologies in educational research. *International Journal of Qualitative Studies in Education,* 30(2), 111–127.

Ceder, S. (2019). *Towards a posthuman theory of educational relationality.* New York, NY: Routledge.

Connelly, F.M., & Clandinin, D.J. (1990). Stories of experience and narrative inquiry. *Educational Researcher,* 19(4), 2–14.

Dasberg, L. (1983). Pedagogy in the shadow of the year 2000. *Phenomenology + Pedagogy,* 1(2), 117 – 126.

De Dalai Lama (1999). *Wijsheid voor een moderne wereld – Ethiek voor een nieuwe tijd.* (P. de Bakker, J. van der Meer, W. Oostendorp, & E. Segeren, Vert.). 'S-Gravenhage: Uitgeverij BZZTôH.

Dohmen, J. (2007). *Tegen de onverschilligheid: Pleidooi voor een moderne levenskunst.* Amsterdam: Ambo.

Edwards, R. (2012). Theory matters: Representation and experimentation in education. *Educational Philosophy and Theory,* 44(5), 522–534.

Elwick, S. (2020). Merleau-Ponty's 'wild Being': Tangling with the entanglements of research with the very young. *Educational Philosophy and Theory,* 52(2), 149–158.

Freire, P. (1972). *Pedagogy of the oppressed.* Harmondsworth: Penguin.

Freire, P. (1995). *Pedagogy of hope: Reliving pedagogy of the oppressed* (R.R. Barr, Trans.). New York, NY: Continuum.

Hannis, M. (2015). The virtues of acknowledged ecological dependence: sustainability, autonomy and human flourishing. *Environmental Values,* 24(2), 145–164.

Hart, T. (2014). *The integrative mind: Transformative education for a world on fire.* Lanham, MD: Rowman & Littlefield.

Hodder, I. (2014). The entanglements of humans and things: A long-term view. *New Literary History,* 45(1), 19–36.

Ingold, T. (2008). Bindings against boundaries: Entanglements of life in an open world. *Environment and Planning A: Economy and Space,* 40(8), 1–15.

Ingold, T. (2010). Bringing things to life: Creative entanglements in a world of materials. *ESRC National Centre for Research Methods,* Realities Working Paper 15.

Kroger, J., & Marcia, J. E. (2011). The identity statuses: Origins, meanings, and interpretations. In Schwarts, S.J., Luyckx, K., & Vignoles, V. L. (Ed.), *Handbook of identity theory and research* (p. 31–53). New York, NY: Springer.

Lenz Taguchi, H. (2012). A diffractive and Deleuzian approach to analysing interview data. *Feminist Theory,* 13(3), 265–281.

Lotz-Sisitka, H. (2017). Education and the common good. In Jickling, B. & Sterling, S. (Ed.), *Post-sustainability and environmental education: Remaking education for the future* (p. 63–78). London: Palgrave MacMillan.

MacIntyre, A. (1990). The privatization of good: An inaugural lecture. *The Review of Politics*, 52(3), 344–377.

MacIntyre, A. (1999). *Dependent rational animals: Why human beings need the virtues*. London: Duckworth.

MacIntyre, A. (2007). *After virtue: A study in moral theory* (3rd edition). Notre Dame, IN: University of Notre Dame Press.

Mazzei, L.A. (2014). Beyond an easy sense: A diffractive analysis. *Qualitative Inquiry*, 20(6), 742–746.

Pikhala, P. (2018). Eco-anxiety, tragedy, and hope: Psychological and spiritual dimensions of climate change. *Zygon*, 53(2), 545–569.

Polanyi, M. (2009). *The tacit dimension*. Chicago, IL: University of Chicago Press.

Ricard, M. (2015). *Altruïsme – De kracht van compassie*. (M. Stoltenkamp, Vert.). Utrecht: Uitgeverij Ten Have.

Rosa, H. (2017). Dynamic stabilization, the Triple A Approach to the good life, and the resonance conception. *Questions de communication*, 31, 437–456.

Son, C., Hegde, S., Smith, A., Wang, X., & Sasangohar, F. (2020). Effects of COVID-19 on college students' mental health in the United States: Interview survey study. *Journal of Medical Internet Research*, 22, 1–14.

Wahl, D. C. (2016). *Designing regenerative cultures*. Axminster: Triarchy Press.

Walburg, V. (2014). Burnout among high school students: A literature review. *Children and Youth Services Review*, 42, 28–33.

On Teaching the Entangled Student

In Chap. 4, I explored three axiological perspectives that might inspire a pedagogical response to students' entanglement in contemporary societal challenges: (1) entanglement-orientedness, (2) entanglement-awareness, and (3) hopeful action. In this chapter, I shift focus to the question of praxeology. Specifically: how can a teacher invite entanglement-orientedness (n.b. and thus challenge students to avoid self- and other-neglect), contribute to the deepening of entanglement-awareness, and support students to act from a position of hope? This chapter aims to articulate perspectives that help teachers in this cause. In doing so, my intention is not to suggest that there is one right way that all should follow; as emphasized before, the complexity of education demands situating the question of effectiveness within the particular relationality of unique teaching situations (Akkerman et al. 2021; Kincheloe and Tobin 2015). For this reason, I proposed, in Chap. 1, to understand helpful perspectives as hermeneutic lenses that can inspire and enable teachers to explore their own practice and articulate their own situated insights and identify new possibilities. To articulate such perspectives in this chapter, I continue the diffractive methodology described in Chap. 4 (Barad 2007; Ceder 2019; Lenz Taguchi 2012; Mazzei 2014); I depart from the praxeological insights articulated by co-researchers as presented in Chap. 3 and read these through selected complexivistic pedagogical theory. As the pedagogical dimensions of

axiology and praxeology are interrelated, the scholarly work I engage with in doing so overlaps to a large extent with that of Chap. 4. In addition, important sources of inspiration for this chapter are the theorizing of Sam Crowell and David Reid-Marr on the process of emergent teaching and the work of Parker J. Palmer and Geert Kelchtermans on the selfhood and integrity of teachers. These scholars have been a source of inspiration since the early stages of my research and, so I wish to suggest, can be considered true pioneers in exploring, carefully, the deep implications of embracing complexity as a teacher.

As the first step in this chapter, I will explore the pattern of insight – as it emerged in inquiry with co-researchers – that in pedagogy of entanglement myriad different teacher moves can be efficient as long as they are attuned to the process of inquiry within entangled phenomena. In doing so, my engagement with, especially, Sam Crowell and David Reid-Marr lead me to understand this process as a cyclical process of opening, organizing, and consolidating inquiry. Reconnecting to Karen Barad, Simon Ceder, and Paolo Freire helps me, subsequently, to interpret this process through the lens of relational ontology (e.g. opening, organizing and consolidating as relational phenomena, the lack of full control on behalf of the teacher, the diffractive nature of inquiry). Next, I move on to a more in-depth consideration of how the open-ended, emergent nature of inquiry within entangled phenomena calls upon a

© The Author(s), under exclusive license to Springer Nature Switzerland AG 2022
K. R. Wessels, *Pedagogy of Entanglement*, Sustainable Development Goals Series,
https://doi.org/10.1007/978-3-031-15787-5_5

teacher to try and do "the right thing here and now" even though s/he can never fully know what that means beforehand, or perhaps never at all. My starting point for this exploration is the set of insights articulated by co-researchers in diffractive step 2.2. Reading these insights, primarily, through the theorizing of Sam Crowell and David Reid-Marr, Parker J. Palmer, Alasdair MacIntyre, Karen Barad, and Geert Kelchtermans, I come to suggest that the art of teaching the entangled student lies in an ongoing practice of *perceptiveness* and *integrity*.

Inquiry Within Entangled Phenomena

The praxeological insights articulated by co-researchers in diffractive step 1.2 suggest that there is not one ultimate teacher move that characterizes pedagogy of entanglement. Teachers, rather, perform a multitude of different moves that, to quote Fig. 3.3, 'can be suitable in a given moment'. Harvested in diffractive step 1.2 are such moves as creating a moment of silence for reflection, following students' initiatives, letting students' initiatives meet resistance and friction, inviting students to speak up, inviting or introducing other or different voices, sharing your own entangledness with students, and providing structure. All such moves, enacted at the right time, can help students gain entanglement-awareness and shape hopeful action, and the list could be expanded. It appears, in other words, that the primary pattern of insight generated in diffractive step 1.2 is not that a certain teacher move is central in pedagogy of entanglement, but rather that teacher moves are attuned to a process of collaborative inquiry within entangled phenomena. A related idea articulated in Chap. 4, is to consider entanglement-orientedness not as an accomplishment that can be achieved once and for all by an individual student, but rather as what Biesta refers to as 'a never-resolved existential challenge' (Biesta 2020, p. 97). Pedagogy of entanglement – as I consider it – seeks to continually increase our awareness of our dynamic entanglement in a world-in-becoming and invites

us to shape and reshape our commitments to the world's becoming. Inquiry is, then, a dynamic, open-ended process. There is no defined endpoint, a moment in time at which a student has become fully aware of his/her entangledness and has once and for all resolved the challenge of shaping his/her response to the challenges of the world. There is only the process of embracing and practicing one's existence as co-shaper of the world, the process of, in Freire's words (1972, p. 52), 'the action and reflection of men and women upon their world in order to transform it'.

Pedagogy of entanglement, in this line of reasoning, is to be understood as a process-centered pedagogy, meaning: its primary focus is not on precisely determined outcomes and fixed learning trajectories, but rather on the collaborative endeavor of discovering relationality and acting within it. For this reason, I have been inspired by Crowell and Reid-Marr (2013), whose work on emergent teaching explores how through responding to what arises naturally in the educational process, we can collaboratively connect with 'the fact that we are part of the environment, within it, moving dynamically with it' (p. 1). Perhaps, their insights can help us frame the process of inquiry. In fact, Crowell and Reid-Marr propose a simple cyclical model that might come to our aid. According to them, namely, emergent teaching 'opens, inquires, and consolidates only to recursively repeat that process' (p. 49). If we look at one cycle of emergent teaching, this looks like the following. First, a certain event draws attention to a specific question or theme and provides 'the loose boundaries and the uncomfortable uncertainty that is needed' (p. 47). Once an opening is there, has attracted attention, inquiry can follow: we can engage in a way which 'organizes the possibilities around relevant questions, the need for new information, and personal interests' (p. 47). Through this inquiry, insights and actions emerge, and as this happens the need for consolidation arises, which 'narrows, understands, and seeks to apply these to what is perceived as needed or what seems most important' (p. 47). Consolidation works, in a way, like closure, it helps to understand, structure, and apply what has been done and learned. Yet it is also a prelude for a new opening, for what we consolidate is not only insight but also intentions and new questions that invite further inquiry. Crowell and Reid-Marr call emergent teaching, therefore, 'a continuous process of opening and closing', and state that 'each time the process is opened up it adds another layer of complexity' (p. 48).

Similar models to that of Crowell and Reid-Marr are recognizable in the work of several of the other scholars I am engaged with. For instance, Palmer (2017) speaks of 'the process of knowing' (p. 117) as (1) gathering around a subject, (2) asking good questions (i.e. not too open or big and not too closed or small) and listening attentively, and (3) reframing (i.e. 'to articulate what we have learned in a way that relates it to where we have been and where we are about to go', p. 138). Lengkeek (2016), to give a second example, in his attempt to identify ways to support students as co-shapers of the world, proposes a process of (1) eliciting students' own actions, (2) channeling students' actions (i.e. not too complex nor too simple and neither too wide nor too narrow), and (3) sharing and evaluating. There is a clear parallel between these models, not only in the three phases but also in the underlying intention to approach students as entangled in the process of understanding and shaping our shared world. The cyclical model of opening-organizing-consolidating inquiry can provide, then, a promising perspective for teachers seeking to contribute to the emergence of entanglement-awareness and hopeful action.

In the paragraphs to come, I will explore this model through the lens of entanglement. In doing so, I shall reconnect with the three diffractive scripts 'The multicultural classroom', 'The sustainable school', and 'Mock and prejudice'. For the sake of exemplarity, and to illustrate how one kind of teacher move can be effective in different parts of inquiry, I shall herein focus on the teacher move 'sharing your own entangledness with students'. In my inquiry with co-researchers, this particular move surfaced frequently and this is, I think, not surprising. For, the insight of entanglement concerns the teacher just as much as it does the student. The contents of inquiry are not some sort of material external to and mastered by the teacher. A teacher may embody more experience and richer entanglement-awareness than his/her students concerning the matter of interest, but, to quote Freire once again (1972, p. 63): 'at the point of encounter there are neither utter ignoramuses nor perfect sages; there are only people who are attempting, together, to learn more than they now know'. Bringing into remembrance the criticism of representationalism (Barad 2007; Polanyi 2009; see Chap. 3), we ought also to recognize that a teacher's approach to a subject is always, inevitably, colored by his/her biography. In pedagogy of entanglement, this is not a problem but a source for inquiry and learning. It is in recognition hereof, also, that Palmer (2017) stresses that as teachers, most fundamentally, 'we teach who we are' (p. 2). As teachers, it is neither possible nor desirable to completely hide our particular entangledness in the world from our students. The challenge is to share it intentionally, in such a way that it contributes to the collaborative learning process. How, then, can we recognize the teacher move of sharing your entangledness in the three diffractive scripts across the phases of opening, organizing, and consolidating inquiry?

In the diffractive scripts 'The sustainable school' and 'Mock and prejudice', the teacher move of sharing your entangledness plays a clear opening role. In the former, it is the fact that this particular teacher is known in the school as a sustainability advocate that triggers the two students to approach him/her; had this teacher not been open about his/her passion for sustainability, the initiative of the sustainability committee might not have been taken, or at least it would have been less likely to take off. In the latter, it was the teacher's choice to present mocking cartoons about Donald Trump and to express his/her disapproval of Trump, that triggered the Trump-supporting student's response, and thus opened the opportunity for inquiry. These two examples interestingly show that sharing aspects of your entangledness as a teacher can open up inquiry in different ways. Students may recognize something in the expressions of their teacher, which can inspire and activate them. Yet it can also happen, as in the Trump example, that students are frustrated by their teacher's expressions, that they experience or think something very different, and are triggered to contrast the teacher. Both responses – inspiration and frustration, alignment and contrast – can fuel inquiry.

The teacher move of sharing your own entangledness is also recognizable in the diffractive scripts as a tool to organize inquiry. In the script 'The multicultural classroom', the opening is constituted by the student's remark 'I bet those f*cking Moroccans did it', which gives rise to a sudden change of energy in the classroom. The teacher feels this as well and is clearly affected. Considering what to do, the teacher decides to share his/her discomfort, saying 'I am not sure how you feel right now, but what "A" just said, as a joke or out of anger, affected me. It makes me feel uncomfortable and I feel that I cannot just continue as if nothing happened. Do you feel something similar?'. In this case, the move of sharing your own entangledness thus works as an attempt to start and structure inquiry. It functions as an invitation to other students to join the conversation and as an example that it is safe to speak up. Something similar happens in the diffractive script 'Mock and prejudice'. Here, the teacher decides to acknowledge to the students that the Trump-supporting student's comment made him/her aware of his/her own prejudice, and the teacher uses this to organize a collaborative inquiry. In both these examples, thus, it is the teacher's ability to be vulnerable, to appear in front of students as imperfect, that is crucial in turning the opening into a collaborative, organized inquiry. The energy is very different, again, in the script 'The sustainable school', for in this script (i.e. in scene 2) the students and their teacher share their enthusiasm with each other and get carried away in a creative imagination of possibilities for a sustainability committee. In this case, the sharing of the teacher contributes to a feeling of recognition, inspiration, and belief in a collaborative project.

Sharing your own entangledness as a teacher, lastly, is also visible in the phase of consolidation in the diffractive script 'The sustainable school'. At the end of the academic year, a meeting is organized for sharing and performances. Perhaps the more common thing to do, on this occasion,

would be to preserve the stage for students – for is it not all about them? – yet here the teacher takes the stage to share insights and experiences as well. In the consolidation phase, sharing your own entangledness as a teacher can be an acknowledgment of the fact that teachers are learners too, and this can deepen the learning process for all who are involved. In the other two diffractive scripts, the phase of consolidation is not quite achieved yet. Both these scripts end with the realization that the lesson is almost over yet the inquiry is unfinished and unconsolidated, and in both cases, the teacher announces plans to continue the inquiry in the next lesson.

The three phases of opening, organizing, and consolidating inquiry can provide a framework, as the above examples illustrate, to understand how particular teacher moves can be effective in pedagogy of entanglement. A similar analysis could be made for other teacher moves, yet I consider the more important point to emphasize here the process of opening, organizing, and consolidating inquiry itself. Many moves can enable this process, depending on the context and actors involved, yet it is the process that is primary. As I understand it here, an opening is an event that brings to our shared attention a certain theme, challenge, or question that permeates our lives in the world, and triggers an experience of opportunity, energy, perhaps even urgency, to engage. Following an opening, inquiry can be organized; we can collaboratively share, seek, and contemplate different perspectives and experiences and experiment with our active involvement within the phenomena of interest. As we do so, the need for consolidation arises: the need to harvest and organize entanglement-awareness, (re)shape our commitments and actions, and perhaps articulate new questions or challenges so that the process of inquiry can carry on. In light of the premise of entanglement, I move on to suggest next, there are some important points to highlight considering the nature of this process of inquiry.

First of all, I find it important to consider the opening, organization, and consolidation of inquiry as relational phenomena (Barad 2007) rather than actions of singular individuals. In the diffractive script 'The multicultural classroom', for instance, the opening does not so much reside in the student who makes the statement 'I bet those f*cking Moroccans did it', but in the relationality of this student, the article they are reading, the particular classroom dynamics at that moment in time, perhaps certain previous experiences of this student with Moroccans, and so forth. It is 'a certain gathering together of threads of life' (Ingold 2010, p. 4) that gives rise to the opening, that accumulates in this student's particular action. The same logic applies to the organization and consolidation of insights, which rather than as a matter of an individual person analyzing and concluding from a distance can be understood as the emergence and ordering of insights through collaborative engagement. Consequently, as a teacher you 'can influence but not control what is happening' (Crowell and Reid-Marr 2013, p. 1); teacher, students, and the subject/material/others they engage with are all co-shapers of inquiry, a process, then, which is inherently open-ended and shared. I agree, therefore, with Lengkeek (2016) that the dynamics of eliciting students' actions can be such that you, as a teacher, take the initiative to create an inspiring situation that elicits a creative response but just as well that you follow and support an initiative taken by a student or, I might add, another form of life we are in touch with.

A related point to make, is that inquiry can be both enabled and frustrated by structure. To stay with the script 'The multicultural classroom', we can consider, for instance, that the student's Moroccan-statement is enabled by the fact that students are gathered in a room for this particular lesson, but that simultaneously this statement frustrates the smooth progression of the teacher's predefined plans for that lesson. To embark on the path of inquiry, then, sometimes requires a teacher to break free from an imposed structure (e.g. to deviate from a schedule), yet the moment s/he does so, the need to structure immediately rises again; for inquiry to reach its potential, it needs to be organized around shared questions, methods, time and space (Crowell and Reid-Marr 2013; Palmer 2017). It supports inquiry, thus understood, if the organization of education is not overly rigid, if there is sufficient space within

curricula to improvise and to take initiative (Crowell and Reid-Marr 2013).

Furthermore, following the theorizing of such scholars as Barad (2007), Ceder (2019), Freire (1972), and Crowell and Reid-Marr (2013), it is important to understand inquiry as a process that occurs *within*, and contributing to, the becoming of the world. This point goes to the heart of the move from an ontology of individualism to one of relationality and has important implications. If we look at the opening of an inquiry, for instance, we could say from the perspective of an ontology of individualism that an opening is like an opportunity to become a participant, to be introduced to a certain theme or question in the world and acquire a seat at the table, so to say. From the perspective of a relational ontology, however, the seat at the table is inherently ours; to live is to be a co-shaper of the world and although there may be endless aspects about my entangledness in the world I am unaware of, the process of entangled learning is not one of entering the world but one of learning about and transforming participation in the world. We all have access to the dynamics of the world, and thus deserve to be taken seriously, or as Freire puts it (1972, p. 37): 'we are all knowledgeable, for we are immersed in the world'.

Following this line of reasoning, the dynamics of inquiry are, in line with Barad (2007), not those of representation – i.e. of trying to learn from some sort of outside position – but those of diffraction, of intra-acting within the world. An important function of teacher moves is, then, to invite and facilitate particular intra-actions within a particular thematic context. Teacher moves can orchestrate (1) students' revisiting of earlier experiences through the lens of a new insight or perspective, (2) dialogue in which students explore their diverse experiences and insights together, (3) the encounter with a certain piece of news, art, literature, philosophy, science, or some other material that is particularly interesting given the inquiry at hand (4) or the intra-action with certain human or non-human others who play a particular role that is relevant to the theme or question of interest. As teachers, we might see it as our task to ensure that when we inquire within a certain theme, a plurality of voices is heard, which is crucial if we aim for entangled flourishing. Only through an open and critical engagement with a plurality of voices can we collaboratively find ways to co-create a world that is shared better than it is today. As co-researchers Pien, Ronald, and Sandra conclude in diffractive step 1.2, the teacher somewhat resembles a choir director: 'you provide instructions so that everyone can make him/herself heard and together you create music'.

Lastly, if we understand inquiry thus as inquiry *within* entangled phenomena, it has the potential to embody both the qualities of realism and resistance. It can be realistic, first of all, in the sense that through inquiry we can become increasingly aware of how the world, and our lives within it, are relationally constituted and what, accordingly, is currently within the realm of the possible (Kuntz 2020). Inquiry can provide us, thus, with what we might call a reality check; in the words of Biesta (2017, p.16): 'when we encounter resistance we could say that the world is trying to tell us something'. Yet pedagogy of entanglement is not just about listening to, and reconciling with, the world as it is, it is also about embracing our roles as co-shapers of the world. By inviting and supporting what I have framed as hopeful actions (i.e. acts of conservation, adaptation, and regeneration) we can aspire to teach students to transgress against destructive forces within our shared world (Freire 1972; Hooks 1994) and thus 'eschew the possible in favor of a challenging potential' (Kuntz 2020, p. 28). Without the quality of realism, inquiry becomes detached from worldly challenges, limits, and possibilities, and therefore cannot be expected to have any real impact, yet without the quality of resistance inquiry becomes docile, devoid of passion, and shall merely reproduce status quo. Inquiry as a process of realistic resistance, however, can catalyze positive change.

Practicing Perceptiveness and Integrity

The recurring cycle of opening, organizing, and consolidating inquiry within entangled phenomena offers a helpful framework to shape pedagogy of entanglement. The course of this process depends not just on the teacher, but on the relationality of all partners involved and has, consequently, emergent and open-ended characteristics (Barad 2007; Crowell and Reid-Marr 2013). Teaching the entangled student implies, then, the ongoing challenge to respond to the particularities of the process of inquiry as it unfolds, such as the sudden Moroccan-statement in the diffractive script 'The multicultural classroom', the resistance of the Trump-supporting student in the script 'Mock and prejudice' and the student initiative for a sustainability committee in the script 'The sustainable school'. Notably, the teacher is him/herself part of the particularity of any given educational moment. In fact, diffractive step 2.1 illustrates that when the different co-researchers reread the diffractive scripts they co-created, many diverse ideas emerged for how these scripts can be improved (see, for the full overview, Chap. 3). Interestingly, for each co-researcher the particular ideas for improvement s/he emphasized drew attention to a particular creative tension between teacher moves. In a given situation, for instance: should one share his/her own entanglement or conceal it?; should one follow a student's initiative or resist it?; should one moderate or spectate?; should one zoom in or zoom out? Diffractive step 2.1 also illustrates that such experienced tensions, in turn, resonate with specific other teaching experiences of these co-researchers. Every co-researcher, in other words, relates to these diffractive scripts in a particular manner, for every teacher is entangled in the phenomenon of teaching through his or her particular biography. This suggests that there is no one simple answer to the question of how one should respond to, for instance, the Moroccan-statement in the script 'The multicultural classroom'. Even if we would be able to keep all other factors exactly the same, an educational situation becomes fundamentally different once we replace one teacher with another. To enact a fitting move once confronted with a challenging educational situation is, in short, a very personal challenge (Kelchtermans 2009; Palmer 2017), that ultimately no one can solve for you. Since every educational moment is unique, we may as teachers become more and more experienced over time but we shall nevertheless be repeatedly confronted with the challenge to shape a response that is effective given the particularities of the here and now. Doing so is, therefore, a dynamic teacher quality that has to be practiced and re-established over and over again. The practice of what kind of quality is this? What does it ask of a teacher to shape his/her unique actions in entanglement-oriented teaching? In the remainder of this chapter, I shall explore two perspectives that might help at this point. To do so, I shall reconnect with diffractive step 2.2 of the inquiry with co-researchers.

Practicing Perceptiveness

When I challenged co-researchers to articulate insights about what it takes to work with a particular tension that emerges in entanglement-oriented teaching, they almost without exception started with emphasizing the importance of "perceiving the situation". Trying to be present in the moment, to be aware of what is going on, to sense, scan, signal, feel, read, notice the situation as it is. For without such awareness, how can one shape any intentional response? To align your teaching to the process of inquiry within entangled phenomena you should, so is the idea, try to be aware of the important particularities of that process as it unfolds. Such perceptiveness is, also, visible in the three diffractive scripts. In fact, in each script an entire scene is devoted to describing what a teacher perceives and how this affects his/her teaching (i.e. scene 3 in each of the diffractive scripts). The importance of perceptiveness is also highlighted in much of the literature I am engaged with. For instance, (1) Crowell and Reid-Marr (2013, p. 128) argue that 'to occupy the space of uncertainty' a teacher should be 'focused and present, open and receptive', (2) Kelchtermans (2009) emphasizes that teaching is an inherently vulnerable profession, in which one has no full control over outcomes yet continuously has to judge and act based on 'an interpretative reading of the situation' (p. 264), (3) Palmer (2017) emphasizes that 'if I want to teach

well in the face of my students' fears, I need to see clearly and steadily the fear that is in their hearts' (p. 46), and (4) MacIntyre (1999) argues that attentiveness to particularities is crucial to the kind of responsiveness that characterizes good teachers.

The perspective of practicing perceptiveness is, then, important for teachers intending to support inquiry within entangled phenomena. Yet what is it that teachers should try to perceive? The insights of co-researchers highlighted, on this matter, five different foci of attention, namely (1) individual students, (2) the group of students as a whole, (3) oneself, (4) society at large, and (5) curricular constraints and opportunities. To provide an example, in the diffractive script 'The multicultural classroom' this could include such elements like: "(1) the student who makes the Moroccan-statement seems frustrated, s/he might have had certain bad experiences that triggered his/her statement, and the Moroccan students in the classroom remain passive, appearing untouched, this is clearly not the first time they hear a comment like this, (2) I notice a somewhat uncomfortable tension in the group, some giggling here and there, it feels like most students are waiting for a response of me or the Moroccan students, and although I think some students are annoyed by the statement it seems to me that right now in this group there is not really a sense of safety and openness to speak up, (3) I notice a conflict in myself, for on the one hand I feel angry and annoyed by this student's statement, and feel the impulse to get angry, but I also recognize the thought that indeed it is quite possible that the criminal is Moroccan because of crime-statistics in the Netherlands, (4) what happens here in the classroom right now is actually a recurring phenomenon in this school and in society at large and links to real problems with integration, social injustice, poverty and criminality, and in fact the other day I read an article that social inequality increased during the COVID-pandemic, and (5) I have about 20 minutes left on the clock and another lesson with them in two days, in which I could follow up, yet we also have a test soon which asks attention".

There are, of course, many more elements or details that could be added to this example, and yet this list is already, at least to me, dazzling in complexity. Seen as individual, separate elements, it is very challenging, if not impossible, to process all of this simultaneously. Perhaps, then, the lens of entanglement can help to get some grip. Seen through this lens, to distinguish five different foci of attention is to perform an agential cut (Barad 2007), that is to say: if we consider these different entities and their impact in isolation, we analytically "cut" the relational threads that bind them together. It is, in other words, not in their isolated individuality but their dynamic relationality that students, teacher, school, and society at large constitute the potential for inquiry. We can draw several interrelated conclusions pursuing this line of thought. To start, the very entanglement-awareness we seek to co-create through inquiry appears to be the same awareness a teacher aims for through practicing perceptiveness. This interestingly shows that the teacher is, fundamentally, part of the inquiry. It is only insofar that the teacher inquires him/herself, and thus continually learns, that s/he can teach within entangled phenomena. This also means, then, that the teacher need not perceive everything perfectly, but rather should be attentive to those things that are essential to his/her task (Endsley 2000), the task of facilitating and enriching the opening, organization, and consolidation of collaborative inquiry. To ask a teacher to be fully aware of the outcomes of inquiry prior to it, would be to re-establish a strict teacher-student dichotomy that. It would be an attempt to put the teacher in an impossible outsider position of absolute knowledge, and that, from the perspective of relational/complexivistic thinking, would only harm inquiry.

As a teacher, then, it is much more important to notice relational patterns that provide an opportunity or obstruction for inquiry, than to be fully aware of every detail there is to know. Such opportunities and obstructions, notably, can be both topical (e.g. the opportunity to inquire into the theme of multiculturalism) and cultural (e.g. a classroom culture of fearfulness can inhibit collective inquiry). Regarding the latter, it is worth noting that co-researchers (see, especially, diffractive step 2.2) emphasized the importance of a culture of sufficient safety and mutual trust and that such scholars as Crowell and Reid-Marr (2013), Rosa (2017), and Palmer (2017) underscore that this is crucial for, respectively, emergent teaching, experiences of resonance, and the experience of connectedness and meaning. The point, here, is not that education should be a fearless process, in which everyone always feels completely safe and trusts everyone limitless, but that if we feel chronically unsupported, unacknowledged, hopeless, or mistreated, we are likely to shut down rather than open up, to hide rather than inquire. As Palmer frames it (2017, p. 40): 'the fear that shuts down the capacity for connectedness is often at work in our students. If we could see that fact clearly and consistently – and learn to address our students' fears rather than exploit them – we would move toward better teaching'.

Practicing perceptiveness as I understand it here, then, is not about noticing everything perfectly, but about trying to become aware of relational dynamics – in which the topical and cultural are interwoven – that provide opportunities or obstructions for collaborative inquiry. In the example of the diffractive script 'The multicultural classroom', the teacher need not know exactly what makes the student utter the Moroccan-statement or why the Moroccan students and other classmates remain silent, but it is crucial that s/he senses that this particular relational dynamic might be important to address as (1) it is deeply rooted in today's multicultural society, (2) students keep their emotions and frustrations to themselves, perhaps out of a lack of trust or safety, and (3) the risk of alienation lures. As a teacher, you may have certain ideas about why students do what they do (e.g. ideas about social-economic background, the influence of parents, defense strategies, fears, etc.) but it is important to be aware that these ideas are not facts but rather interpretations that can be improved. If teachers are, thus, inquirers themselves, it is inevitable that they are sometimes wrong. As a teacher, it is, therefore, crucial to allow your interpretations to be challenged, and if necessary transformed, by the responses you receive. For instance, maybe the Moroccan-statement is much more an attempt to make a provocative joke, out of boredom, than an actual expression of anger and frustration toward Moroccans. If this is so, the above-mentioned interpretation might be exaggerated and it would perhaps make more sense to just move on or to focus inquiry on the experience of boredom and how we respond to it. In this sense, the luck of the teacher is that students are part of the inquiry as well. Once the teacher takes the initiative to shape inquiry around the theme of multiculturalism, the student who uttered the comment might for instance say something like 'don't take it so seriously, I have many Moroccan friends myself, I'm just teasing', and the teacher might, in turn, interpret this a cue to ask the student 'why did you want to tease?', thereby leading the conversation in another direction. The point is, thus, that as a teacher the purpose of practicing perceptiveness is not to be fully aware and completely right, the purpose is to continually learn and respond, as part of the attempt to contribute to collaborative inquiry. The perceptive teacher need not be perfect or enlightened, s/he needs to be curious, present, and open to learn.

Practicing Integrity

Equally important to the practice of perceiving as such, the practice of being aware of what is important in the here and now, is the practice of responding to what you perceive in a way that contributes to the process of collaborative inquiry. I have found it fruitful to frame this practice in terms of integrity. I do so inspired by my co-researchers and the theorizing, in particular, of Palmer (2017). In describing how they utilize their perceptiveness to make decisions in challenging educational situations, my co-researchers regularly used the word "integrity" as an overarching concept. When I, in such a case, would ask what they meant by this, one or several of the interpretations harvested in diffractive step 2.2 would emerge, the interpretations of (1) commitment, (2) authentic style, (3) professional real-ism, (4) constructive self-awareness, (5) collegial support, and (6) professional independence. This led me to suspect that integrity might be a promising, rich perspective (i.e. with multiple interrelated interpretations) to use in the context of pedagogy of entanglement.

The term "integrity" appealed to me particularly for it refers both to the quality of being true to moral principles and to the state of being whole, undivided. To integrate, notably, is the act of bringing together in harmony. This makes integrity an interesting concept in the context of relational ontology and pedagogy of entanglement. The moral principles we intend to be true to in pedagogy of entanglement – i.e. those of entanglement-orientedness – advocate an integral approach to the good life, one in which the flourishing of a plurality of lives is integrated within the flourishing of community. Such, also, is how Palmer (2017) introduces the concept of integrity as an essential teacher quality. Integrity, for Palmer (p. 14), requires that 'I choose life-giving ways of relating to the forces that converge within me […] ways that bring me wholeness and life rather than fragmentation and death'. These forces that converge are, in Palmer's terminology, 'the inner and outer forces that make me who I am, converging in the irreducible mystery of being human' (p. 14). Integrity, then, is not a quality someone can possess, as an essence, but a quality, rather, that can be ascribed to how one responds in a particular situation. Take, for instance, the interpretation of integrity as the practice of an authentic style (i.e. integrity as trying to stay close to your passions and to ways of working that work for you, as also experienced by your students); this authentic style is not some stable, unalterable characteristic but a dynamic quality that evolves (Barad 2007; Ceder 2019; Palmer 2017). What stays, is the attempt to do something in a way that resonates with what you believe in and makes you feel confident, joyful, and effective. Integrity, in other words, is a process rather than an essence and something to practice rather than something to possess. In the paragraphs to come, I will read into the six interrelated interpretations of practicing integrity as they emerged in the inquiry with co-researchers.

Whereas in Chap. 4 and the current chapter thus far I focused on the scripts 'The sustainable school' and 'The multicultural school' respectively, I will now focus on the script 'Mock and prejudice' for exemplarity. Together, these six interpretations of integrity can support teachers' ability to transform perceptiveness into teacher moves that facilitate and enrich the opening, organization, and consolidation of collaborative inquiry within entangled phenomena.

The first interpretation of integrity is that of *commitment*. In this perspective, to practice integrity entails trying to have a clear embodied pedagogical intention and to be congruent in basing your actions hereon. To be committed, thus understood, implies both intention and act; it implies first of all to have a purpose, something to strive for, and second of all a dedication to that purpose, as visible in what one does and how one responds (Fournier et al. 2020; Novacek and Lazarus 1990). In Palmer's formulation (2017, p. 6), teaching is an intentional act of creating conditions that can inspire students on 'an inner journey toward more truthful ways of seeing and being in the world'. Building forth on Chap. 4, in pedagogy of entanglement this commitment might be similarly articulated as follows: to facilitate and enrich the collaborative inquiry through which we become increasingly aware of our entangledness in the world-in-motion and practice the hopeful embrace of our roles as co-shapers of the world. To practice integrity, means to let this commitment run through your heart, mind, and hands, to experience it, to try to live up to it. Yet, this is still quite abstract. In specific educational situations, what a teacher needs is not just such a general commitment, but situated sub-commitments, based on situated judgments (Kelchtermans 2009; MacIntyre 1999). For instance, in the diffractive script 'Mock and prejudice' the strive for entanglement-awareness and hopeful action becomes centered around the theme of prejudice, and the teacher's decision to openly admit to having blind spots of him/herself is rooted in the suspicion that it is challenging for students to be open and vulnerable about this. In other words, the rationale behind the teacher move of sharing your own entangledness is, in

this case, to create a context in which open dialogue is possible, and this, in turn, contributes to the higher purpose of increasing entanglement-awareness and inviting hopeful actions.

Following these considerations, commitment truly is a practice; the question always is: if such and such is my higher purpose, what is important to do here and now? Notably, the higher purpose is itself a matter of debate. I have, in Chap. 4, developed an argument for how we might consider the axiology of pedagogy of entanglement, but I do not have the illusion that the matter is, thereby, resolved. Purposes evolve, are dynamic, plural, and embodied, and we do good to keep questioning them. This is especially relevant if we realize that what we perceive is mediated by our commitments (Endsley 2000; Kelchtermans 2009). Would the teacher in the script 'Mock and prejudice' solely be committed to transferring a predefined set of knowledge within this particular lesson, s/he might not have noticed the rich opportunity for inquiry once the Trump-supporting student spoke up, and would likely just have continued without giving it much attention. But even though we might criticize the legitimacy of the commitments of this teacher, s/he would at least have been driven by a clear purpose and can be considered to demonstrate more integrity than the teacher who has no lively sense of purpose at all. In that scenario, randomness and volatility lure, and how can we trust someone whose responses are all over the place? As co-researchers emphasized, teaching amidst complexity does not necessarily have to make you feel disoriented and out of control. The more you experience and commit to a lively sense of purpose, the easier it is to navigate unknown terrain, to experience a sense of agency and identity, and to be transparent in your actions for students and colleagues (Kroger and Marcia 2011; Wahl 2016).

The second interpretation of integrity is the practice of an *authentic style*: the effort to stay close to your own passions and to ways of working that work for you, as also experienced by your students. Whereas commitment is more about being true to a pedagogical cause and the educational process aimed at that cause, authentic style is about being true to oneself in that effort. Having worked closely with my co-researchers, I have come to see them all as unique teachers, and they present themselves as such. For instance, Jens presents himself as a teacher who likes to joke and who prefers a lot of freedom and uncertainty. He is in his element when he is only half prepared and when he can let his enthusiasm and sense of humor lead him into lively discussions with his students. Kasper, on the other hand, presents himself as a calmer and more structured teacher. He prepares for lessons carefully, feels more at ease in one-on-one interactions than in group discussions, seems to be easily trusted by his students, and feels that he needs to process slowly and deeply to come up with good pedagogical responses. I could, of course, give more examples, but the point is that it does not make sense to expect different teachers, like Kasper and Jens, to perform their work in identical manners. They excel, and feel comfortable, in different ways. Of course, as emphasized, their authentic styles are not fixed; Kasper may discover the improviser in himself and Jens might come to value a side of himself that is more structured and meditative, and both, notably, might become better teachers by doing so. The point is, however, that if Kasper and Jens are not allowed to teach in a way that feels personal to them, of their own making, they are likely to be both less effective and less appreciated by their students (Kelchtermans 2009; Palmer 2017). Palmer articulates the point sharply when he concludes (p. 11):

> 'From years of asking students to tell me about their good teachers […] it becomes impossible to claim that all good teachers use similar techniques: some lecture nonstop and others speak very little; some stay close to their material and others loose the imagination; some teach with the carrot and others with the stick. But in every story I have heard, good teachers share one trait: a strong sense of personal identity infuses their work.'

Several of my co-researchers emphasized, in agreement with Palmer, that students are very capable of discerning which teacher infuses his/her work with a strong sense of personal identity and which teacher is merely playing a role, and that it is the former that inspires them. From the perspective of entanglement, this makes perfect sense; the teacher who infuses his/her teaching with a personal style is an example of embracing one's role as co-shaper of the world. Rather than hiding behind a mask s/he says: this is me, who are you?

The third interpretation of integrity – *professional realism* – refers, first of all, to the practice of protecting your fitness for the job by not crossing your limits (e.g. emotionally, physically, idealistically). The interpretation of professional realism, like that of an authentic style, emphasizes the importance of being true to yourself. Perhaps, for instance, the teacher in the diffractive script 'Mock and prejudice' does not feel strong enough on that particular day to embark on the path of an open-ended group discussion about such a sensitive theme as the prejudices we hold. Maybe s/he is very tired, is disturbed by family issues at home, or is otherwise distracted. One could say: so what, as a professional you should be able to ignore all that and just do your job. But, from the perspective of entanglement we ought to recognize that it is both extremely difficult, if not impossible, and in the long term self-destructive – and thereby destructive for the educational process – to cross your limits over and over again (Palmer 2017). Doing the right thing with what you perceive, thus, also means not asking more of yourself than you can currently handle. The best decision a teacher can make in any particular situation, in short, is not some ideal decision that any teacher should make in that situation, but the best *that* teacher can do in *that* particular situation. Integrity, thus, also entails practicing self-compassion and realism, not being too hard on yourself and not expecting the impossible of yourself, but appreciating your professionalism and your performance as it evolves.

The second part of the perspective of professional realism refers to the practice of putting your role as a teacher in a realistic perspective which neither underestimates what you have to offer due to life experience nor overestimates the reach of your influence and knowledge. There are, thus, two potential pitfalls here. The first is that a teacher believes or aspires to have more power and control over students and their learning than s/he actually has and/or expects to know all answers to the questions we encounter in the process of inquiry. In Chap. 2, we have seen that amidst complexity this is an unrealistic expectation and such scholars as Crowell and Reid-Marr (2013), Freire (1972), and Palmer (2017) indeed emphasize that teachers are, in fact, also learners. Crowell and Reid-Marr, notably, emphasize that whilst 'we are conditioned to think we have to have all the answers and that any disruption is bad' (p. 128), to be successful in creative adaptation teachers need to embrace the space of uncertainty and need to realize that 'they can influence but not control what is happening' (p. 1). The teacher in the diffractive script 'Mock and prejudice' is a beautiful example in this sense: during the educational process, s/he becomes aware of his/her own prejudice and takes this not as a fatal injury imposed on his/her authority and legitimacy, but rather as an opportunity to learn and facilitate collaborative inquiry. Again, then, integrity entails not being too hard in judging your shortcomings as a teacher, to allow yourself to be imperfect and to learn. Yet the opposite pitfall also lures. Being aware of our imperfection, and the limits to our influence, we might, as teachers, conclude that we have nothing to teach, that our role is simply to facilitate students learning but not to color that learning in any way with our unique contributions. I agree with Biesta (2017) and Palmer (2017) that such a student-centered approach is to be avoided, for it rejects the student the opportunity to learn from the life experience and knowledge of the teacher. The conclusion of the limits to the power and knowledge of the teacher ought not to be the end of the teacher who teaches, but rather a sense of what Freire (1972) calls humility. It is insofar as a teacher participates as a co-learner, and thus speaks from his/her own experience and is open to new perspectives, that s/he can teach valuable life lessons to his/her students. The authority to teach is granted, as Palmer (2017, p. 34) frames it, to those who are perceived as 'authoring their own words, their own actions, their own lives, rather than playing a scripted role at great remove from their own hearts'. To practice integrity, thus

seen, includes the effort to sincerely participate in collaborative inquiry with students, and thereby to neither impose your truths on students as universal facts nor withhold them altogether but rather to share them as living material to engage with.

The fourth interpretation of integrity that emerged through inquiry with co-researchers, is that of practicing *constructive self-awareness*: trying to be aware of your own entangledness with thematics at hand and the intervention you tend to enact as a matter of habit or reflex and to utilize this awareness in a way that serves collaborative inquiry. This interpretation of integrity emphasizes that, just like students, teachers are entangled within the phenomena of inquiry (Barad 2007; Ceder 2019; Palmer 2017). Consequently, as a teacher, you sometimes find yourself particularly touched by something that happens, as it resonates with your biography intensely. Such moments, which following Kelchtermans (2009) we could call moments of vulnerability, surfaced frequently in my conversations with co-researchers, and two particular examples from co-researcher Astrid got stuck in my memory. These two experiences, namely, interestingly contrast each other. In the first experience, a student of Astrid spoke up during a lesson to strongly plea in favor of the tradition of "black Pete" (i.e. the black-skinned helper of Saint Nicolas in the yearly "Sinterklaas"-festivities in the Netherlands), and in the second experience a student spoke up during a lesson about the Second World War saying 'well, the fact that so many Russians died during the war saves a lot of CO-2 emissions today'. Both the themes of black Pete and the Holocaust resonate strongly with Astrid's biography, and she embodies strong opinions on them that oppose those articulated by these students. In these two particular situations, however, she reacted in very different ways. In the first example, she lost her temper and clashed with the student in front of all other students, leading to an apology on her behalf later on and a feeling of regret and shame. In the second example, she felt the same impulse but managed to inhibit herself and entered into a more open dialogue. Through this, so Astrid experienced, the student came to realize the impact of his statement – especially on people like Astrid whose family history involves loss in the Second World War – and Astrid managed to include the other students in this valuable insight on the often unintended impact of our words. These two particular examples show that, in such moments of vulnerability, roughly two things can happen: (1) you can be overwhelmed and uncontrolled in your reactions, or (2) you can remain able to process your emotions and thoughts in that moment and to shape your response with a degree of agency and a pedagogical intention. It is the latter reaction that is an impressive manifestation of integrity on behalf of the teacher.

This particular interpretation of integrity emphasizes that if we connect to our own emotions, thoughts, and experiences with a sense of curiosity and calmth, if we manage to welcome them without being overwhelmed, we have the unique opportunity to speak with an authentic, intimate, non-violent voice that can truly touch students and deepen collaborative inquiry. Doing so, hard as it is sometimes, is a success story of both being true to yourself – you take your own emotions and thoughts seriously – and being true to your students and a pedagogical cause: you transform your experience of being touched into a response that enables rather than obstructs collaborative inquiry. As Palmer (2017, p. 2–3) observes, 'teaching holds a mirror to the soul', and 'if I am willing to look in that mirror and not run from what I see, I have a chance to gain self-knowledge – and knowing myself is as crucial to good teaching as knowing my students and my subjects'. In this sense, the teacher's effort in the diffractive script 'Mock and prejudice' is also admirable: in the moment s/he is confronted by a student s/he manages to not only allow the confrontation, but to process it, to become aware that indeed s/he taught from a perhaps somewhat biased position, and manages, also, not to get angry or offended or overly insecure but rather to utilize this moment pedagogically by opening up collaborative inquiry.

The fifth interpretation of practicing integrity is that of *collegial support*: trying to keep critically questioning and developing your pedagogical views and actions together with colleagues and other educational professionals. Especially considering that as teachers our interpretations and responses are always mediated by our particular entangledness in the world-in-motion (Barad 2007; Polanyi 2009), we must allow them to be contrasted by those of our professional peers, who have similar concerns and experiences. In the dialogue with colleagues, we are called upon to make our judgments explicit, so that 'others can comment, question, elaborate, contradict and thus contribute to furthering its validity' (Kelchtermans 2009, p. 264). As MacIntyre (1999) emphasized, in our ability to make our own situated, practical decisions we depend on others, and in the context of a particular profession 'we have no one else to rely on but those who are our expert coworkers, to make us aware both of our particular mistakes in this or that particular activity and of the sources of these mistakes' (p. 95).

As teachers, we thus depend in our practice of integrity on our colleagues and on the school and educational system we are part of. Particularly, are teachers supported in their authenticity, and is there a culture of learning and experimenting together? It is for these reasons, also, that Palmer (2017) pleas for teachers to gather in community in 'a space centered on the great thing called teaching and learning' (p. 164). The point Kelchtermans, MacIntyre, and Palmer try to make is, notably, not that all teachers should think and act the same, but that to validate and improve our teaching, and to feel confident once we encounter a challenging situation, we are immensely helped by the inspiration, feedback, and confirmation of expert-others. It is, in fact, in contrast to these others that we can recognize certain strengths, weaknesses, and personal preferences in our actions. Sincere participation in collaborative inquiry within the phenomenon of teaching, thus, shows that one is not trapped inside his/her own bubble of thought, but is open to learn and improve, and that, so is to be clear, is an essential part of the practice of integrity.

The sixth interpretation, lastly, is that of integrity as practicing *professional independence*, meaning: trying to nurture the courage, calmth, and flexibility to make and justify your own choices in the moment, even if these diverge from initial plans and expectations. This interpretation highlights that, although it is crucial to be in touch with colleagues, in the end you have to make your own situated decisions whilst teaching. Looking at the diffractive script 'Mock and prejudice' we might imagine at this point a teacher who lively experiences the purpose to organize a collaborative inquiry into the prejudices we hold, has an idea of how to do so, and is willing to share aspects of his/her own prejudices to this end, yet nevertheless holds back because s/he is scared to fail in the eyes of colleagues, students, or students' parents. Just as it is part of the practice of integrity to be open to feedback, it is also important to have the courage to trust your situated, professional judgments and enact them. Only then, if you allow yourself the authority (Palmer 2017) to teach, does it become fruitful to partake in the professional dialogue with colleagues. Only then is this very dialogue, and thus a professional community, possible, for what point is there in deliberating our actions if we do not dare to act based on our judgments?

The very point of collegial support, as respectively Crowell and Reid-Marr (2013) and MacIntyre (1999) consider it, is to get comfortable with uncertainty and emergence, and to strengthen the ability to make situated decisions. Integrity, in this sense, thus means both to sincerely engage with others in the ongoing dialogue aimed at questioning, validating, and improving your pedagogical interpretations and actions, and to have the courage to make up your own mind, to try and experiment. To consider, thus, the challenge of doing the right thing as one that depends on the freedom to make your own decisions within the professional community you are part of. Practicing collegial support and professional independence, together, make a teacher trustworthy, for although s/he will undoubtedly be imperfect, at least we can trust that s/he will not get lost either in his/her own bubble of unchallenged believes or in the web of expectations and believes imposed by the educational system and colleagues. To practice integrity in

this sense truly is challenging and needs courage (Palmer 2017). It is, however, a practice worth committing to. It is, in fact, the very same practice we hope, in pedagogy of entanglement, to seduce our students to commit to, the practice of embracing one's unique role as co-shaper of the world-in-motion whilst being open and sensitive to the needs, responses, and initiatives of the others we are entangled with.

At the beginning of this exploration of the perspective of integrity, I emphasized that it is to be understood as a process and practice rather than as an essence and possession. I might add, now, that throughout this chapter integrity has emerged as a multi-directional concept. The interpretations of commitment and constructive self-awareness emphasize integrity as being true to a pedagogical cause, thus focusing attention on the educational process aimed at this cause. The interpretation of constructive self-awareness also emphasizes, along with those of an authentic style and professional realism, that in doing so it is crucial to be true to yourself: to respect your limits, to have a healthy, realistic sense of your task and influence, to allow yourself to infuse your work with a personal style, and to remain open to learn about yourself. The two interpretations of collegial support and professional independence, lastly, emphasized the importance of being true to the larger professional community of educational professionals, by both nurturing the courage to make and enact situated judgments and openly sharing and questioning them in dialogue with expert-others.

Engaging with these interwoven interpretations of integrity can help a teacher to shape his/her responses amidst challenging educational situations. They provide legitimate criteria to navigate within complexity so that an answer to the question "what should I do" can emerge. Being true to a pedagogical cause, as we have seen, helps to prioritize what is important and what is not, and forms a basis for the ability to adapt and improvise in the here and now. Being true to oneself, on the other hand, provides a teacher with a more practical navigation tool. It helps to decide how ambitious to be, which methods to choose, and what to do with one's own emotions and thoughts. Based on this personal dimension of integrity, different teachers might respond differently to similar challenges, and it is exactly because of these differences that they might each enjoy their work and be successful. Being true to the professional community of teachers, lastly, helps a teacher to remain open to learn, continue to strengthen and justify his/her practical reasoning, and nurture the courage, confidence, and willingness to try new things that the quality of teaching so greatly depends on. Simultaneously, practicing integrity can help to provide a teacher with a sense of trustworthiness and honesty in the eyes of students, colleagues, and parents. The teacher who practices integrity is likely, as Palmer (2017) states, to receive the authority to teach. His/her commitment, honesty, and openness invite students to engage likewise, and is that not his/her deepest hope? If we truly aspire to develop complexity-embracing pedagogical responses to contemporary societal challenges, we should therefore not only revisit curricula to create more opportunities for collaborative inquiry. We need also, or perhaps most of all, to support and inspire teachers in the ongoing process of practicing perceptiveness and integrity.

References

Akkerman, S. F., Bakker, A., & Penuel, W. R. (2021). Relevance of educational research: An ontological conceptualization. *Educational Researcher,* 50(6), 416–424.

Barad, K. (2007). *Meeting the universe halfway: Quantum physics and the entanglement of matter and meaning.* Durham: Duke University Press.

Biesta, G. J. J. (2017). *The rediscovery of teaching.* New York, NY: Routledge.

Biesta, G. (2020). Risking ourselves in education: Qualification, socialization, and subjectification revisited. *Educational Theory,* 70(1), 89–104.

Ceder, S. (2019). *Towards a posthuman theory of educational relationality.* New York, NY: Routledge.

Crowell, S., & Reid-Marr, D. (2013). *Emergent teaching: A path of creativity, significance, and transformation.* Lanham, MD: Rowman & Littlefield.

Endsley, M. R. (2000). Theoretical underpinnings of situation awareness: A critical review. In Endsley, M. R., & Garland, D. J. (Ed.), *Situation awareness analysis and measurement* (p. 3–32). Mahwah, NJ: Lawrence Erlbaum Associates.

Fournier, V., Bretonnière, S., & Spranzi, M. (2020). Empirical research in clinical ethics: The 'committed researcher' approach. Bioethics, 34(7), 719–726.

Freire, P. (1972). *Pedagogy of the oppressed.* Harmondsworth: Penguin.

Hooks, B. (1994). *Teaching to transgress: Education as the practice of freedom.* New York, NY: Routledge.

Ingold, T. (2010). Bringing things to life: Creative entanglements in a world of materials. *ESRC National Centre for Research Methods,* Realities Working Paper 15.

Kelchtermans, G. (2009). Who I am in how I teach is the message: self-understanding, vulnerability and reflection. *Teachers and Teaching: Theory and Practice,* 15(2), 257–272.

Kincheloe, J. L., & Tobin, K. (2015). Doing educational research in a complex world. In Tobin, K., & Steinberg, S. R. (Ed.), *Doing educational research: A handbook* (2nd ed.) (p. 3–13). Rotterdam: Sense Publishers.

Kroger, J., & Marcia, J. E. (2011). The identity statuses: Origins, meanings, and interpretations. In Schwarts, S.J., Luyckx, K., & Vignoles, V. L. (Ed.), *Handbook of identity theory and research* (p. 31–53). New York, NY: Springer.

Kuntz, A. M. (2020). Resistance is becoming not possible: Philosophical inquiry and the challenge of material change. In Denzin, N. K., & Giardina, M. D. (Ed.), *Qualitative inquiry and the politics of resistance: Possibilities, performances, and praxis* (p. 24–39). New York, NY: Routledge.

Lengkeek, G. (2016). *Pedagogisch leiderschap: het ondersteunen van vorming door onderwijs in exacte vakken.* Delft: Uitgeverij Eburon.

Lenz Taguchi, H. (2012). A diffractive and Deleuzian approach to analysing interview data. *Feminist Theory,* 13(3), 265–281.

MacIntyre, A. (1999). *Dependent rational animals: Why human beings need the virtues.* London: Duckworth.

Mazzei, L.A. (2014). Beyond an easy sense: A diffractive analysis. *Qualitative Inquiry,* 20(6), 742–746.

Novacek, J., & Lazarus, R. S. (1990). The structure of personal commitments. *Journal of Personality,* 58(4), 693–715.

Palmer, P. J. (2017). *The courage to teach: Exploring the inner landscape of a teacher's life* (20th anniversary ed.). Hoboken, NJ: Jossey-Bass.

Polanyi, M. (2009). *The tacit dimension.* Chicago, IL: University of Chicago Press.

Rosa, H. (2017). Dynamic stabilization, the Triple A Approach to the good life, and the resonance conception. *Questions de communication,* 31, 437–456.

Wahl, D. C. (2016). *Designing regenerative cultures.* Axminster: Triarchy Press.

Living the Question of Integrity

Having explored helpful axiological and praxeological perspectives for teaching the entangled student throughout Chaps. 3, 4, and 5, I will devote this chapter to an autoethnographic engagement with the perspective of integrity as developed in Chap. 5. I do so with two purposes in mind. My first purpose is to make it easier for teachers to engage in the practice of integrity. My intention, after all, has been to articulate helpful perspectives which enable them to reach their own situated conclusions. The easier it is for teachers to engage personally with the outcomes of educational research, the better. It is in great part for this reason, also, that I have tried to enrich my writing with exemplary experiences of co-researchers. Chapter 3, for instance, provides short narratives illustrating how co-researchers try and make deliberate choices in challenging teaching situations (see diffractive step 2.1), and I hope that reading these triggers reflections of your own. In this chapter, I hope to further contribute to the "relatability" of the six interpretations of integrity by presenting how I have been engaged with them myself through autoethnography. It is worth noting, in this context, that as a researcher I have been in quite a comparable position to that of the teacher in pedagogy of entanglement. To be more precise, I have led my co-researchers through three cycles of opening-organizing-consolidating inquiry within the complex question of teaching the entangled student, as Table 6.1 illustrates. An autoethnographic account of my practice of integrity throughout the research process with my co-researchers is, then, exemplary in two ways: it provides an example of how the method of autoethnography can enable one to engage systematically with the six interpretations of integrity, and it provides an example of what kind of insights such an engagement might bring about in a pedagogical context.

My second purpose is to further enrich my thesis with the quality of transparency (Ali-Khan and White 2019; Ellis et al. 2011; Taylor and Settelmaier 2003). As I introduced in Chap. 1, for me this whole inquiry has been an intrinsically motivated quest rooted in my own experiences and ambitions as an educational change agent, and throughout it, I have attempted to nurture a congruence between emerging theory and methodology. In line with this complicity (Davis and Sumara 2006), I must consider the outcomes of the inquiry not as material external to myself but as perspectives that concern me personally. As soon as I suggest, in other words, that it would be valuable for teachers to take the practice of integrity very seriously, I need to do so in my own work as well. I hope, therefore, that reading this chapter provides you with a sense of how my own learning experiences are interwoven with the research findings I have presented and leaves you with the impression that my efforts have been sincere and committed to a clear, pedagogical cause.

K. R. Wessels, *Pedagogy of Entanglement*, Sustainable Development Goals Series,
https://doi.org/10.1007/978-3-031-15787-5_6

Table 6.1 The inquiry with co-researchers summarized in three cycles

	Opening	Organizing	Consolidating
Cycle 1	Letter of invitation describing my motivations, research questions, and research plans, and discussing these at the start of narrative biographical interview	Semi-structured narrative biographical interview with individual co-researchers exploring their biographies and exemplary teaching experiences	Co-researchers make summarizing notes during interview sessions and I construct and communicatively validate narrative synthesis texts afterward
Cycle 2	30-min phone call with each co-researcher to discuss the premise of entanglement and my plans for a group session; co-researchers read biographical narratives of each other	Group assignment with 3 or 4 co-researchers to co-construct a challenging diffractive script exemplary for pedagogy of entanglement	Sharing insights in groups regarding the question 'how can I work meaningfully with the entangledness of my students' and collaboratively organizing them into mind-map structures
Cycle 3	Letter of invitation describing my interest in revisiting the results of cycle 2 and plans for another 1-on-1 session with each co-researcher; co-researchers reread the co-constructed diffractive scripts	Co-researchers engage in five individual sub-assignments aimed at rewriting diffractive scripts, identifying pedagogical tensions, and extrapolating these to recent or current teaching experiences	Collaboratively summarizing axiological and praxeological insights during 1-on-1 sessions; I construct and communicatively validate a summary of all harvested insights afterward

An Autoethnographic Engagement with the Six Interpretations of Integrity

Having stated my purposes in this chapter, I move on, now, to my methods. To structure my autoethnographic account, I have first of all reformulated the six interpretations of integrity into six questions. These questions, so is the idea, are questions to live (Wahl 2016), questions to keep asking oneself and each other in order to continually learn. These questions are, in the context of this thesis, aimed at teachers or educational professionals in general, but could be used in other professional contexts as well. The questions are:

1. Commitment: to what extent do you experience a lively sense of purpose in your work, how would you describe it, and how do you succeed and/or fail to let it direct your actions?
2. Authentic style: to what extent, and in what kind of situations, do you experience in your work that you are in your element, true to your passions and strengths?
3. Professional realism: how do you experience the challenge to ask neither too little nor too much of yourself in your work, and to both embrace your responsibility and influence and accept their limits?

4. Constructive self-awareness: how is your sense of self touched or challenged in your work, and how do you succeed and/or fail to utilize this experience for the sake of collaborative inquiry?
5. Collegial support: to what extent, and how, do you sincerely question and develop your views and actions in dialogue with expert others?
6. Professional independence: to what extent, and in what kind of situations, do you succeed to make and enact your own situated judgments?

Autoethnography – writing about experiences and insights that emerge while being in a particular position or context – combines the qualities of autobiography (i.e. writing about your own life as you experience it) and ethnography (i.e. writing about the culture you are participating in) (Ellis et al. 2011). To enable this process, as also explained in Chap. 1, throughout my inquiry I have utilized the following three strategies: (1) I have collected personal notes of triggering events, insights, thoughts, and emotions that I experienced throughout the inquiry, (2) I have written yearly process reports depicting the experiences that were most important to me and what they taught me, and I have shared and discussed these reports with senior colleagues, and (3) at the end of the inquiry with co-researchers I gathered feedback regarding how they had experienced the research process and my role in it. It is not difficult to imagine that these three strategies, in one form or another, can be – and in varying degrees they already are – utilized in a teaching context. A teacher might, for instance, keep some sort of professional diary, periodically discuss reflections and experiences with colleagues or superiors, and frequently ask for specific feedback from students and colleagues.

The data generated by following these three strategies enable me to provide answers to the six questions above that provide an impression of how my sense of self evolved and of the kind of insights that can emerge through the practice of integrity. It is, nevertheless, important to acknowledge the limitations of my answers in a double sense. First of all, even though they are also informed by the voices of relevant others, they are not objective or final. Yet, the aim of practicing integrity, and of autoethnography to this end, is not to arrive at objective, final knowledge, but to be transparent while learning and to deepen self-awareness. Secondly, such autoethnographic accounts are inevitably incomplete. It would be neither interesting for a reader or listener nor possible for a narrator – me in this case – to reveal everything experienced and learned. My strategy here, as is common in autoethnography (Ellis et al. 2011), is therefore to focus on those experiences and insights that for me are most striking and transformative. I now move on, then, to taking on the questions I have posed. I hope that you find it stimulating to read my interpretations and, most of all, that doing so inspires your own practice of integrity.

Commitment: To What Extent Do You Experience a Lively Sense of Purpose in Your Work, How Would You Describe It, and How Do You Succeed and/or Fail to Let It Direct Your Actions?

I trust that reading this book conveys that I am, indeed, driven by a lively sense of purpose. In the process report I wrote in 2019, 1 year after starting this book-project, I summarized my core concern as the following question: if we embrace complexity, what is a viable pedagogical narrative to work with? As Chap. 1 illustrates, I consider this question of immense importance both on a personal and a systemic/societal level. Personally, I have grown up with an intense desire to be included in, rather than excluded from, the existential and societal challenges of living in today's world. It has taken me almost 30 years to come to understand that, in fact, I am and always have been included, and that it is our common language and the way we educate that tends to suggest otherwise. On a systemic/societal level, it worries me to see that although there is a growing awareness, in academia and society, of the complexity and interconnectedness of contemporary challenges, this is insufficiently manifested in the way we educate young people. How can we expect ourselves and fellow citizens to be wise and responsible in the face of complexity and interconnectedness when our educational systems still approach the world from a distance, as a fixed, stable entity to be prepared for and introduced into by following a linear, predetermined path?

Although I have been driven, thus, by a sense of purpose and urgency, I have also experienced that the discipline and patience to do what is needed in light of that purpose is quite another thing. Allow me to provide two short examples hereof. In the early stages of my research, I realized that it was important to consider my own experiences during this research project as valuable data to enrich inquiry. For this reason, as also highlighted before, I decided to collect personal notes of triggering events, insights, thoughts, and emotions that I experienced throughout the inquiry. Yet, although I am and have been convinced of the value of doing this, I found it hard to consistently do so. I have tried different strategies, such as carrying a notebook with me to take daily notes, taking notes on my phone or laptop, or regularly writing short articles or columns. I do experience that I have collected, taken together, a rich set of experiences that helped me to develop my thesis, but it cost me a considerable amount of mental energy to bring up the discipline and I would be lying if I said that this discipline has been consistently high. I have experienced a similar struggle in processing the data of sessions with co-researchers. In order to, in the terms of Polanyi (2009), dwell in the data, I found it important to do all the processing myself, rather than to outsource the tasks of transcribing, summarizing, and communicatively validating data. To complete the hours and hours of concentrated work needed for these tasks has been, indeed, a contest with myself, and I have had numerous frustrating days in which I discovered I am not at all alien to laziness and procrastination. In short, it tends to come quite natural to me to experience a sense of purpose and urgency, and to do the tasks that I experience as fun and stimulating (e.g. the actual sessions with co-researchers, discussions with supervisors, reading literature, writing), but to consistently do those things that I know to be important but experience as boring and energy-consuming, is truly a demanding practice. That is what I experience as the main challenge in integrity understood as practicing commitment: to be disciplined in completing important tasks which I do not enjoy as activities in themselves.

Authentic Style: To What Extent, and in What Kind of Situations, Do You Experience in Your Work that You Are in Your Element, True to Your Passions and Strengths?

I started this research project after several years of engaging in experimental educational projects at The Bildung Academy. I did so as I felt a desire to understand more fundamentally what kind of change I was trying to bring about and why so. And, also, to bring together different experiences and thoughts that were somewhat chaotically co-existing in my mind into a coherent narrative. Doing so, seeing patterns and connections, framing the story that singular experiences tell together, is a desire I have always felt, be it as a kid philosophizing with friends, as a student, or in my work. To think systemically, as Wahl

(2016) calls it, is something I enjoy and seem to thrive at, and throughout this inquiry my desire to do so has been truly satisfied, be it in reading, writing, or collaborating with co-researchers. I experienced also, and I had not experienced this so clearly before, that others can benefit from my thinking and insights. For instance, the feedback I received from co-researchers indicates that they were impressed by the accuracy of summaries of insights that I would provide during or after sessions, and that it helped them to articulate sharply and to come to new or deeper insights themselves. For me, it was a real accomplishment to experience that I could use my strengths, thus, in a way that improved the quality of collaborative inquiry. I also became aware, however, that in this quality resides a potential weakness. I experienced this especially in the phase of inquiry in which co-researchers collaboratively constructed diffractive scripts and harvested insights. After these sessions, I was very satisfied because I experienced that we co-constructed rich narratives and organized rich insights. Yet, later I learned that a few co-researchers had felt rather disoriented throughout the process. I felt I had not been sufficiently aware of this although as the facilitator of the process it should have been one of my main priorities. I reckon I was so occupied with recognizing the patterns in the output we were generating that I lost track of the individual needs of co-researchers in the process of doing so. Such, perhaps, is the nature of our authenticity: it can be both our biggest strength and our pitfall. My integrity, then, depends on the effort to let it be the former and to learn from situations in which it becomes the latter.

Professional Realism: How Do You Experience the Challenge to Ask Neither Too Little Nor Too Much of Yourself in Your Work, and to Both Embrace Your Responsibility and Influence and Their Limits?

Throughout this inquiry, I believe I have become more realistic in how I relate to, and take responsibility for, structure. Perhaps it comes not as a surprise to you, given the focus throughout my thesis on concepts such as emergence, and open-endedness, that I am not particularly fond of very structured environments or meetings. I rather like it when there is a bit of chaos and plenty of room for spontaneity and the unexpected. The initiation of The Bildung Academy, also, is in my experience in great part a story of breaking free from outdated structures. Yet, throughout this inquiry, I have come to revalue structure, be it dynamic, as essential to any effort of collaborative inquiry. Structure, that is, not as a predefined and unalterable frame to fit in, but rather as the ongoing effort to keep track of, and provide an opportunity for, collaborative inquiry. Without the focus of attention, the commitment to a certain experiment or activity, and the organization of insights, nothing really happens. The understanding of Crowell and Reid-Marr (2013) that inquiry is a process that continually needs opening and closing inspired me greatly in this sense. My problem, I now understand, is not with structure itself but with structures that are so rigid that they kill the opportunity for emergence.

I thus came to see that it is naïve to refuse to create structure out of a fear to dictate others and to kill opportunities for creativity and emergence. In my sessions with co-researchers, it happened frequently that a co-researcher requested more structure – a stronger sense of what we were doing – in order to come to creative ideas and new insights. In the beginning, this would annoy me, but slowly I came to see it as a crucial part of my task to provide the right amount of structure for every co-researcher. We all need a degree of structure and as a teacher or researcher or leader in general you are particularly responsible for it. Yet, I believe it is also naive to think that there is one predeterminable, perfect structure that shall bring about the best results. To structure collaborative inquiry, in my experience, is an ongoing effort. Sometimes it is needed to break free from structure to make something new possible, sometimes we need to create more structure to stick together and have a sense of orientation, and sometimes we need to differentiate because one needs more or a different structure than another. To take responsibility for the right amount of

structure has, for me, been a great practice in integrity throughout this inquiry. The feedback I have received from co-researchers does indicate that I have often left the space of inquiry more open than they are used to or, in some cases, feel secure with, but also that they have managed to stay engaged and inspired throughout the process. A piece of feedback that I experienced as a true compliment came from Aafke when she said that I did not direct the outcomes of inquiry but framed the opportunity for it, for that, indeed, is the kind of enabling structure I believe we need more in education.

Constructive Self-awareness: How Is Your Sense of Self Touched or Challenged in Your Work, and How Do You Succeed and/or Fail to Utilize this Experience for the Sake of Collaborative Inquiry?

Above all, this inquiry for me has been a practice in listening to and learning from the perspectives and experiences of others: co-researchers, teachers I meet through my work for The Bildung Academy, and so forth. There is one particular voice that now and then (and luckily less now than before) pops up in my head and threatens my ability to do so. This voice says something like: "the wisdom is inside me, and perhaps a few wise people, all the others are superficial and boring". I have become increasingly aware of this voice in the initial phase of my inquiry when I was shaping my methodology. I noticed, then, a tendency in myself to want to do everything in my own mind, by myself; to read a lot, reflect on my own experiences, build my own arguments, and write. Yet, rationally I was, then, already aware that my research would greatly benefit from a more rich and diverse collection of data providing insight into experiences and perspectives of diverse teachers.

Around the same time, I was following a training program into the methodology of Deep Democracy, which is a methodology for including minority voices into majority decisions and for exploring polarities and contradictions in yourself and groups. This particular polarity – i.e. "I am wise and others are superficial" – surfaced also in this context and I came, then, to several insights. I came to suspect, first of all, how this voice is rooted in my youth. In short, I have been inclined to question why things are the way they are and what is good or meaningful since early childhood, yet only in very few of my childhood peers did I recognize a similar "hunger" (which, notably, does not necessarily mean it was not there). I had one particular friend in my teenage years with whom I used to spend evenings on end debating all kinds of matters and I spent much less time than most of my peers on what I used to see as "superficial partying". Together, perhaps, we created a narrative of "us the intellectual vs them the superficial". There is, so I also came to see, both a truth and a misconception to this narrative. The truth, and on this matter I refer back to the section focusing on integrity as the practice of an authentic style, is that I tend to be, indeed, deeply interested in the big questions of life and that I am talented in seeing and articulating patterns, nuances, and relations. Because of this, I do experience that I am often able to say things that make sense to people and are labeled as wise or inspiring. The misconception, however, is first of all that anything intelligent I might understand or say is "mine" – it emerges relationally and I merely attempt to articulate it – and second of all that others are boring and superficial. We all have our own experiences, strengths, and concerns, and my experience by now is that if I succeed to value these rather than expecting someone else to resemble me, I learn something. I have learned, for instance, a great deal from all my co-researchers, whose teaching experiences and personal reflections have been so central to the development of the ideas presented in this thesis.

To reach my potential as someone who articulates intelligent ideas that can help others, thus, I absolutely need to open myself to the richness of others' thoughts, emotions, and experiences. I feel grateful, therefore, that I received numerous feedbacks from co-researchers describing my role in the inquiry as that of a good listener and an open mind. This confirms my own experience;

for me, namely, the research process with my co-researchers marks the transition of "Koen who wants to solve everything in his own head" to "Koen who engages in dialogue, allows himself to be inspired and helped in his quests by others". This open attitude does not mean, of course, to agree with everyone, nor to consider every argument equally valid, it means simply to acknowledge that we are all entangled in the same world and, therefore, embody experiences, concerns, and perspectives that say something about our shared world and can contribute to collaborative efforts of sense-making. Acknowledging this, and living up to it, is an important challenge in my own practice of integrity, and I intend to continue this practice in the future inquiries I shall participate in.

Collegial Support: To What Extent, and How, Do You Sincerely Question and Develop Your Views and Actions in Dialogue with Expert Others?

I have, to build forth on my previous answer, experienced my collaboration with teachers in the role of co-researchers as great support, and I might add that the feedback and support of my colleagues and participation in collaborative research groups have been crucial to the successful completion of this book. In all these relationships, and on such occasions as conference visits, I have asked of myself to be open to the contribution of others. I want to highlight here, however, how the dynamics of this effort changed throughout the inquiry in line with the emergence of the covid-19 pandemic, which cannot stay unmentioned in this autoethnographic account. I do experience, namely, a sharp contrast in how I organized and valued collegial support before and during the pandemic. Before, I felt part of a very dynamic, lively community. I attended conferences and symposia, and I visited numerous schools and universities through my work for The Bildung Academy. On these regular occasions, I would meet and engage with other educational professionals and I experienced this as a rich opportunity to test out my ideas and gain feedback and inspiration. For instance, my confidence to use the concept of entanglement as the central notion in my work received a strong boost during my visit to the ECER-2019 conference in Hamburg, where I presented my initial ideas for a pedagogical narrative rooted in the insight of entanglement and received various positive reactions and literature recommendations (e.g. the work of Karen Barad). During the pandemic, these offline meetings with new people disappeared, and frankly, I did not feel attracted to attend online conferences for whole days while most of my work already took place behind the computer. What stayed, however, and in some cases intensified, is the regular meeting with small groups of colleagues: feedback sessions with senior colleagues, monthly meetings with the larger research group I am part of to discuss our work and new perspectives, and participation in a few initiatives of acquainted researchers to form an informal think-thank and/or support group. My circle of support, in other words, has become smaller during the pandemic but also intensified.

I reckon that the collegial support we need in order to keep questioning and improving our work changes over time. For my own integrity, I can now say that I find it important to find a balance between the spontaneous, dynamic encounter with new people and the long-term collaboration with teachers and researchers close to me. I do feel that in the second half of my inquiry the balance has shifted to this second type of support, and perhaps it is time to shift it back again a bit, to be more pro-active in meeting new people and benefit from the wide range of opportunities to be supported by the worldwide research community.

Professional Independence: To What Extent, and in What Kind of Situations, Do You Succeed to Make and Enact Your Own Situated Judgments?

I do believe, all in all, that I have made my own situated decisions throughout this inquiry. Such decisions as to work with teachers in the role of co-researchers, to focus on the insight of entanglement, and to use a diffractive methodology, were of course not taken in a vacuum – and this, also, is not what the interpretation of integrity as the practice of professional independence advocates – but no one handed these decisions to me as instructions to follow; I came to my own conclusions about what would benefit inquiry and shaped my actions accordingly. My decisions, notably, have often not been mainstream. In fact, many of my methodological and pedagogical conclusions can be interpreted as criticism of the majority of educational research and practice, and I am aware that this does not always make it easier to be at home in the community of educational professionals. I must, one might think consequently, be very confident about my work, and often I am indeed. However, I must also admit that from time to time I have to deal with what I suspect most researchers and teachers have to deal with: existential self-doubt. As my girlfriend can testify, every now and then I struggle with the fear to turn out to be a fraud, someone who should not be taken seriously for his ideas are based on delusions and misconceptions. I am well aware that if such self-doubt persists too strongly, it becomes impossible to arrive at coherent, constructive situated judgments. It is not a basis for clear thinking, nuance, and insight. I am also aware, however, that often my self-doubt is partly justified. I have experienced enough times, by now, that I disagree with my earlier self, or that colleague's point toward a shortcoming in my work, to know that my theorizing is not, and will never be, perfect. But the strong, existential tone of my self-doubt – "I am delusional!" – is, so I am convinced by far most of the time, unjustified.

My integrity, in my experience, depends a great deal on my ability to feel – rather than to think – that this existential self-doubt has no valid ground, that of course my work is not perfect but neither is it worthless. I have experienced the following factors as crucial in this struggle for the strength, courage, and trust that professional independence depends on: (1) the day to day encouragements of those dear to me, especially my girlfriend, (2) the encouragement and positive feedback of close colleagues and, in a lesser degree, other researchers and educational professionals, and most important of all (3) enough time and space for myself to think freely, to read, to (re)connect to my values, purposes, and inspirations, or in short: to take good care of my intellectual needs.

The fact that you are reading my work, is proof that these three conditions have been sufficiently met. Yet, this struggle for confidence does leave me with one last observation about my own integrity. Taking my responses to this question and the question regarding constructive self-awareness together, it appears to me that I run the risk of neither taking myself nor others seriously. The risk of believing, in other words, that no one has anything worthwhile to say. To avoid such a nihilistic, relativist position is, perhaps, the key challenge for me in practicing integrity. Integrity, for me, is about a dialogical way of being in the world, about taking both myself and others seriously in the collaborative quest for a beautiful, meaningful, flourishing world!

Closing Remarks

When I discussed the autoethnographic account I have just presented with senior colleagues, one of the topics that came up was that answers and questions appear to overlap somewhat. For instance, in describing how I have become more realistic toward the importance of structure, I also talked about how it initially annoyed me when co-researchers requested more structure and how my biography is characterized by a tendency or urge to break free from structure. I could have presented this answer, therefore, also in response to the question of constructive self-awareness. Or, to provide another example, my tendency and desire to ask meta-questions and think systemically surfaces – in slightly different ways – both in response to the questions concerning authentic style and constructive self-awareness. Based on such observations, I would like to close this chapter with some remarks on how to approach the six questions/interpreta-

tions of integrity. The six questions, namely, function merely as a tool to identify that which is currently important or "hot" in the practice of integrity for s/he who engages with them. Engagement with the six questions can help to reach an overall feeling or experience of integrity and meaningful effectiveness, and that, rather than ticking the boxes of six different types of integrity, is what the practice of integrity is all about. Perhaps the struggles and insights that I have shared in my account mean little to you as what I am in the process of learning is self-evident in your experience, or perhaps you resonate strongly with some parts of my writing as you have similar experiences. Perhaps, also, you find one or some of the six questions more interesting or confronting to engage with than others, as you feel or anticipate that they call forward something of particular importance to you. Or maybe you notice a certain pattern in your answers to the different questions as something similar, apparently of importance, keeps popping up. I do not think, therefore, that the six interpretations of integrity are some sort of perfect or final categorization of integrity into six separate and exhaustive domains, but rather that they offer an interwoven set of questions that can continue to trigger personal and professional learning throughout our lives.

References

Ali-Khan, C., & White, J. W. (2019). Between hope and despair: Teacher education in the age of Trump. *Educational Philosophy and Theory, 52*(7), 738–746.

Crowell, S., & Reid-Marr, D. (2013). *Emergent teaching: A path of creativity, significance, and transformation.* Lanham, MD: Rowman & Littlefield.

Davis, B., & Sumara, D. J. (2006). *Complexity and education: Inquiries into learning, teaching, and research.* New York, NY: Routledge.

Ellis, C., Adams, T. E., & Bochner, A. P. (2011). Autoethnography: an overview. *Historical Social Research, 36*(4), 273–290.

Polanyi, M. (2009). *The tacit dimension.* Chicago, IL: University of Chicago Press.

Taylor, P. C., & Settelmaier, E. (2003). Critical autobiographical research for science educators. *Journal of Science Education Japan, 27*(4), 233–244.

Wahl, D. C. (2016). *Designing regenerative cultures.* Axminster: Triarchy Press.

In Conclusion: Pedagogy of Entanglement

I have arrived, now, at the end of my inquiry for as far as I have been in the lucky position to shape it within the context of this book. Bearing in mind Crowell and Reid-Marr's understanding of inquiry as 'a continuous process of opening and closing' (2013, p. 48), I would like to devote this last chapter to a recapitulation of outcomes and a consideration of several paths worth pursuing in moving forward. Let us start with the former. Inspired by Meirieu's model for pedagogical doctrines (2016), Fig. 7.1 provides a summary of the helpful perspectives developed throughout the preceding chapters. Together, these perspectives provide a heuristic intended to invite and help teachers to explore and shape their situated pedagogical responses to the complex challenges that permeate contemporary society.

Let us walk through Fig. 7.1. In doing so, I invite you to take a particular educational opportunity or experience in mind that you would like to explore further through this heuristic.

We start in the center, in the dimension of ontology, where we are invited to explore how challenges and transformations in our shared world are complex. We can recognize, for instance, how challenges we face are multi-interpretable and open-ended, how our actions and their effects are situated in larger systems that are themselves in motion in ways beyond our design, and how both within and between systems and challenges interconnectedness and co-dependence surface. Notably, we are ourselves part of this complexity in motion, and we can explore how every one of us, students and teachers alike, are entangled with complex societal challenges. We can recognize how our identities and positions in the world have been shaped by the particular ways in which complex societal challenges have touched our lives and how we, through our actions and commitments, have had – and currently have – a hand in their evolution. As such, we can come to suspect a fundamental relationality from which the experience of self and other emerge.

Let us now take a next step and formulate the ambition to respond to this complex reality pedagogically. To do so, we need both a sense of purpose and a practical approach. We start with the former and thus move down in Fig. 7.1 to the dimension of axiology. Here, we can build on the insight that the well-being and creative potential of any human being is entangled with that of others, both human and more-than-human. This can move us to transcend an educational focus on either individual accomplishment or collective conformity, and to strive, rather, for entangled flourishing, for a collective sense of us in which individuality is engendered. Rather than competition and separation, we may advocate cooperation and seek to inspire students to nurture resonant relationships that are mutually enabling. We may come to see that there are endless opportunities to invite and practice such an orientation in education, for whatever the complex challenge,

K. R. Wessels, *Pedagogy of Entanglement*, Sustainable Development Goals Series, https://doi.org/10.1007/978-3-031-15787-5_7

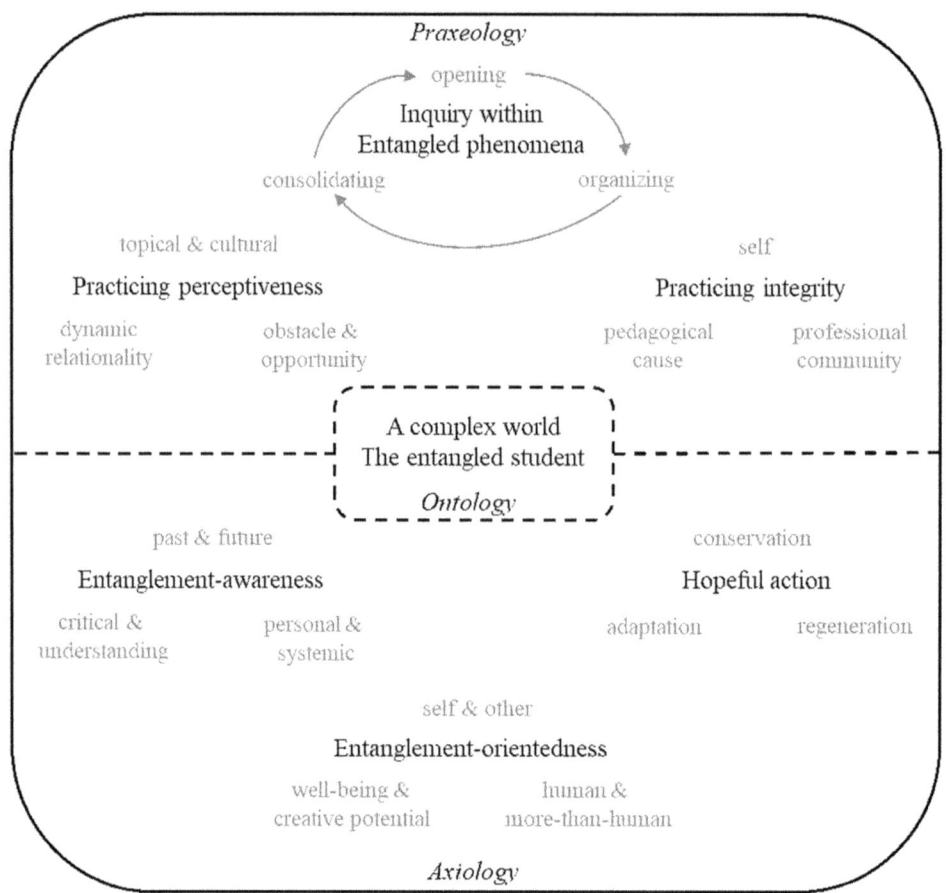

Fig. 7.1 Helpful perspectives for pedagogy of entanglement

the quality of the relational patterns that weave our shared world are at stake.

If we thus move on and seek to invite and practice entanglement-orientedness in education, we may discover that this has at least two dimensions. We can strive, firstly, to invite and support students to become increasingly aware of their dynamic entangledness in the world-in-motion. Together, we can deepen our understanding of our personal paths through the world and become aware of the forces that converge in our unique biographies and the strategies we have developed in response. As we do so, awareness of the larger systems and societal dynamics that our lives are situated in can grow as well. We may pay explicit attention, furthermore, to how entanglement-awareness connects past and future, so that alternative images of the future – which hold a real

potential – may start emerging. Perhaps, as we thus collaboratively deepen our awareness of our entangledness in a relationally constituted world, students discover a critical voice in themselves. As this happens, we may pay extra attention so that the understanding emerges, simultaneously, that however destructive or unfair particular relational patterns may be, they are so for particular reasons that reach beyond simple accountability.

We are, however, not only reflective beings but also co-shapers of the world. In education we may therefore strive, secondly, to invite and support students to express increasing entanglement-awareness through hopeful actions, actions that promote a way of being in the world through which our joint well-being and creative potential can be realized. In any context, we may experience that such actions can take on at least three forms. We may explore and enact ways to conserve existing relational patterns from which mutual flourishing emerges. We may find ways to adapt current relational patterns to reach a better harmony of needs. We may move to regenerate relationships by replacing or transforming destructive relational patterns with/into relationships through which we can win together.

Let us move, now, to the upper part of Fig. 7.1, to the dimension of praxeology. Here we can recognize and experience that as we engage with complex societal challenges in education, there is no one, ultimate teacher move that can be copy-pasted anywhere anytime. What is effective to invite and support entanglement-awareness and hopeful action depends, rather, on the particularities of the here-and-now. We can recognize, nevertheless, certain dynamics of the process of inquiry through which such learning takes place. To do so, we need to be aware that this inquiry – entangled as we are – takes place *within* the world; the potentiality for inquiry lies in the ways in which we are shaped-by and shapers-of complex societal challenges in a here-and-now sense. It may help, therefore, to explore how inquiry can be iteratively opened, organized, and consolidated. To open inquiry, we may seek to recognize or orchestrate events through which students' entangledness with particular complex societal challenges becomes visible. Such events can create a feeling of urgency and engagement and this energy can, subsequently, be canalized into a collaborative inquiry; by staging intra-action with relevant sources, others, or places, by taking on creative/experimental challenges, and by engaging in contemplative dialogue, we can organize conditions from which deeper insight and mindful action can emerge. As this happens, we may start experiencing the need to consolidate; we can seek ways to challenge and help students – and perhaps ourselves as well – to share emerging insights, manifest them in purposeful initiatives, and articulate new questions or intentions that can open up inquiry yet again.

If the iteration of opening, organizing, and consolidating provides a useful framework for inquiry within entangled phenomena, we are called upon to embrace and appreciate its open and emergent character. The teaching profession may present itself to us, consequently, not as a skill to master but rather as an art to practice. It is in a vibrant connection to the particularities of the here-and-now that our moves as teachers can make a difference that matters. To achieve this, it can help, first of all, to engage in the practice of perceptiveness. We may challenge ourselves to be aware of the dynamic relationality of students-teacher-school-society as it evolves. Yet we ought to be careful not to put ourselves in an outsider position as we do so, for we are ourselves part of this relationality. Therefore, rather than expecting ourselves to be fully aware of everything going on, and to hold all the answers, we may see it as our task to sense topical and cultural opportunities and obstructions for collaborative inquiry and to cultivate a curious and open mind for this purpose. We may sense, for instance, how students have diverse, perhaps contrasting experiences concerning a particular theme and see this as an opportunity to learn from each other. Or we may sense, to provide an example of a cultural obstruction, a certain tenseness or unsafety in the classroom that inhibits open dialogue.

Yet the question remains what we are to do with what we perceive, and we may experience that a methodological toolset alone is not enough. It can help, therefore, to furthermore engage in the practice of integrity, the ongoing effort to be true to a pedagogical cause, oneself, and professional community. As we teach the entangled student, we may challenge ourselves to live the questions of integrity, and we may notice how as we do so a general experience of meaningful effectiveness is nurtured. In any situation, we may inquire:

1. What is my pedagogical purpose, and what does it ask of me to do? (Commitment)
2. How can I utilize my personal passions and strengths? (Authentic style)
3. How can I respect my current emotional/physical limits and the lack of full control on my behalf, whilst simultaneously embracing my responsibilities and influence? (Professional realism)

4. How is my sense of self touched or challenged, and how might I utilize this experience to enrich collaborative inquiry? (Constructive self-awareness)
5. How do my professional peers judge this situation, and what does this teach me? (Collegial support)
6. To what extent do I succeed to make and enact my own situated judgment, and what do I need to increase/nurture the courage to do so? (Professional independence)

So, we have come full circle. A pedagogical response to complex societal challenges, so is the argument I have explored, ought to be rooted in the embrace of complexity. Rather than analyzing and attempting to solve complex societal challenges from an outside position, I have become convinced that we need to learn and practice to move within them. Humble and ambitious. Open and critical. Meditative and creative. Let us embrace our roles as dependent co-shapers of the world, and learn together toward a world of mutual flourishing.

New Openings

In this hyperconnected, dynamic world we live in, permeated by profound challenges and transformations, the awareness of complexity is unequivocally on the rise. It is high time that our educational institutions and pedagogical approaches come to embody this growing awareness, to assist and inspire humanity to learn to move within complexity with increasing sensitivity and wisdom. Doing so is necessary, for if there is one thing that the years behind us bear witness of, it is that the tendency and attempt to simplify, separate, control, and indeed exploit has – as the dark side of the advancements of modern life – brought upon us unprecedented ecological and humanitarian crises. For me and my co-researchers, working on this book has been an insightful commitment to the ongoing effort of embracing complexity and learning and acting within it. I truly hope that more and more teachers and educational researchers shall com-

mit themselves likewise and that the heuristic summarized in this chapter can aid this process. I would like to end my contribution, for now, by highlighting five openings to further the cause of a pedagogical response to complex societal challenges.

Educational Research Methodology

Throughout this book, I have advocated the importance of nurturing a resonance between theoretical premises about phenomena of interest and methodological approaches to study these phenomena. Understanding education as a complex process (Biesta 2010, 2016), and positioning the educational researcher within the phenomenon of education rather than outside it (Barad 2007; Polanyi 2009), the following three questions have been implicitly at work: in light of complexity thinking (1) how are we to conceptualize the relation between educational research and educational practice?, (2) what kind of methodologies are promising for which purposes, and (3) how might different methodologies complement and enrich each other to improve academic rigor and quality? In facing these questions, I found myself supported by recent criticisms of the tendency of quasi-causal thinking about (improving) educational practice (especially Akkerman et al. 2021; Biesta 2016). I have, with my choices, attempted to formulate answers to these questions in the context of this book; I have framed the contribution of my research as the articulation and legitimization of helpful perspectives understood as hermeneutic lenses, I have advocated the importance of exemplarity and transparency, and I have opted for interweaving the three methodological approaches of diffractive narrative inquiry with teachers as co-researchers, pragmatic literature study, and autoethnography. My choices, hopefully, can inspire other educational researchers in similar situations, yet there are, of course, more strategies and considerations possible, and I undoubtedly missed several interesting opportunities to further my cause.

Methodologically, it thus seems to me, we find ourselves in challenging and innovative times. My hope for the future of educational research is especially sparked by Kenneth Tobin and Shirley Steinberg's collection of bold visions in educational research (2015). In their introducing chapter, Kincheloe and Tobin (2015) frame the common thread throughout the chapters of the book as the emerging epistemological and ontological realizations that 'knowledge is stripped of its meaning when it stands alone' and that 'to be in the world is to operate in context, in relation to other entities' (p. 5). For educational researchers, they suggest, this gives rise to the need 'to avoid the surge of hyperrationality and the instrumental rationality that characterizes it' yet likewise to resist 'an irrationality characterized by a nihilism and relativism that offer no hope for scholarly growth or ethical action' (p. 5). To do so, so they argue, is not a matter of one methodology for all, but rather a matter of what is increasingly commonly called "bricolage" (see, also, Berry 2015): critically engaging with myriad methodologies and perspectives depending on what is needed and what is at hand. An important and promising path to pursue is, in sum, to recognize, question, and expand emerging methodological insights and approaches in the effort of embracing complexity in educational research.

Teacher Education

If a pedagogical response to the complex challenges that permeate society is something we truly value as an educational community, this cannot but have consequences for how we educate our teachers. Following the inquiry that I have presented here, some logical recommendations would be:

1. Spend considerable energy on studying and discussing complexity thinking in an educational context, especially focusing on such matters as our own and students' entangledness in the world, the situatedness of "what works", the emergent nature of educational processes, and the axiological and ethical implications of teaching in the face of complex societal challenges.

2. Approach the education of teachers itself as a process of collaborative inquiry within the complex phenomenon of education, by welcoming teacher-students as co-researchers with relevant histories and orchestrating rich and diverse possibilities for experimentation and (transdisciplinary) dialogue. The methodological approaches I have utilized throughout this inquiry (i.e. narrative-biographical interviews, collaborative diffractive script-writing, autoethnography, and a diffractive engagement with pedagogical literature) could also contribute to such an approach to teacher education.

3. As part of such an approach to teacher education, emphasize the ongoing practice of perceptiveness and integrity, and thus repetitively challenge teacher-students to approach every educational moment as unique, to shape and reshape their pedagogical commitments and authentic style, to become increasingly aware of their students' and their own situational needs and particular entangledness in thematics at hand, to (dare to) make and enact situated judgments, and to openly share and question their experiences and perspectives together with their teachers and peers.

It would be worthwhile, therefore, to focus inquiry on teacher education and to (1) explore how existing teacher education programs and teacher education research align or contrast with the picture sketched above, and (2) if there might be particularly promising angles for trying to improve teacher education practices. In the context of the Netherlands (and I suspect in other countries as well), there are several trends perceivable that might be particularly interesting to consider in such inquiries. For instance, it is curious to notice new teacher education programs being founded, based on an experienced insufficiency of traditionally available programs, such as the primary school teacher programs "SPRING" (i.e. with a specific focus on innovative schools; Hogeschool Leiden n.d.) and a new 2-year master program focusing particularly on pedagogical challenges in large cities (Erasmus University Rotterdam n.d.). It is also interesting to note that in the Netherlands – and in many other countries as well – we are struggling with a growing shortage of qualified teachers and, consequently, there is a strong call for revitalizing and re-appreciating the teaching profession (Inspectie van het Onderwijs 2019). Might, perhaps, a complexivistic reading of the teacher profession contribute to this agenda? Might the perspective of teaching as a dynamic, emergent process, and the opportunity of helping young people to constructively engage with and contribute to the complex challenges of our times attract candidates who currently opt for other career paths?

The School as a Professional Community

Although the education of teachers is an important way to protect and improve the quality of education, in the context of complexity it is abundantly clear that good teaching is grounded in attentiveness to the here and now and an ongoing process of professional learning and curriculum (re)design. Palmer (2017) translated these insights into his plea for teachers to gather in community in what he called 'a space centered on the great thing called teaching and learning' (p. 164). How, in other words, can we create and nurture a professional culture (e.g. mutual trust, the encouragement and willingness to share, experiment, and provide and receive feedback, a feeling of shared responsibility, etc.) and supportive resources (e.g. time, space, access to literature, facilitation of activities aimed at professional learning and curriculum design, etc.) in schools? The recommendations summarized above concerning teacher education, in fact, can be translated to this context as well, and a promising angle for further inquiry would be to focus on obstacles and opportunities for strengthening collaboration and professional learning of teachers in schools (see, for instance, Fluijt 2018; de Jong 2021).

In the context of a teacher shortage, such as in the Netherlands, it is particularly hard to organize the supportive resources needed for lively professional communities. It is no surprise, therefore, that in the Netherlands teachers often experience excessive work pressure, stress, and a shortage of time for professional- and curricular development (Bijlsma et al. 2017, 2018), and repetitively unite in demanding concrete governmental investments to enable positive change (Stolk et al. 2019). My personal experience, in my work for The Bildung Academy, is, indeed, that teachers often miss time and support to reflect, engage in deep dialogue with colleagues, and critically evaluate and redesign curricula, and that they genuinely appreciate those moments that break this general pattern. This, I believe, should be one of the main points to focus on when trying to improve educational practice.

Curriculum Design

If a school does succeed to organize and nurture a professional community, one of the most exciting potentials, considering complexity thinking, is that this opens up opportunities to collaboratively (re)design curricula and utilize emergent opportunities. Allow me to illustrate this point with an exemplary experience of co-researcher Ronald. Traditionally, Ronald has been one of the lead teachers in organizing and preparing yearly school musicals at his high school. In doing so, he likes to co-create every aspect of the musical together with participating students. Students are involved in theme selection, scriptwriting, decor design, and so forth, and at the end of the year students perform the musical for fellow students, teachers, parents, and other invitees. The most recent experience Ronald worked on with a group of students (i.e. at the moment in time of his narrative-biographical interview), focused on the theme of sexting, which, in today's digital world permeated with privacy concerns and scandals, has become quite a hot topic. This theme, therefore, carries the potential to engage not only the participating students but the school as a whole. Yet, in Ronald's experience, opportunities to fulfill this potential are typically only minimally utilized. In his mind, the buzz created by such a musical project carries the potential to open up the curriculum of various school subjects. We might imagine, for instance, engaging with the script in language courses, debating the ethical dimensions of sexting in philosophy courses, analyzing and discussing the legal dimensions and societal/political initiatives in social studies, exploring the history of social media in history class, working with sexting-statistics in mathematics, creating a safe space for sharing and support in mentor-meetings, and so forth. All such activities, in turn, could feed back into the process of co-creating the musical, thus truly making it a collaborative inquiry of the school as a whole.

From the perspective of complexity thinking, so is to be clear, a curriculum is not a closed, final organization of material to be mastered, but a lively, open attempt to organize and orchestrate our joined efforts to learn within the world. To summarize Kincheloe and Tobin's argument (2015): we have to study the world in context rather than isolating and abstracting objects of study from the interrelationships that give them meaning. Typically, complexivistic readings of curriculum frame it as an open and collaborative process (e.g. Doll Jr. 1993) or conversation (Crowell and Reid-Marr 2013), and interpreted as such the perspective of the iterative process of opening, organizing, and closing inquiry within entangled phenomena is, in fact, a model for curriculum design. It is important to further engage with the question of how curricula can embrace complexity, what obstacles stand in the way of doing so, and how we might overcome such obstacles. In doing so, it is worth noting that, in line with Ronald's experience, Crowell and Reid-Marr observe that 'in the present-day climate of outcome standards, arbitrarily paced content expectations, and one-size-fits-all approaches to instruction, the idea of emergent teaching runs counter to the general mindset prevalent in most educational institutions' (p. 109). Yet, a more open approach to curriculum does not necessarily mean throwing away all outcome standards. Although this book has focused on the need to open up our educational processes to the embrace

of complexity, I do believe that there are valid reasons to explore how the logic of preparation can find its place within a more holistic and open approach to education (see Chap. 2). I consider it very promising to pursue and explore, therefore, the idea that part of the art of teaching is to creatively connect collaboratively agreed upon outcome standards to emergent teaching processes. In all the examples provided above on how a school musical can open up curricula of school subjects, in other words, every teacher can engage with such emergent opportunities with his/her own curricular agenda in mind (e.g. "I want my students to learn grammar rule X", "I want to teach my students how to critically evaluate data sources and information", etc.).

Exemplary Pedagogical Practices

My core focus in this book has been to articulate pedagogical perspectives that can help teachers to explore and shape their relationship with students in the face of complex societal challenges. In the preceding paragraphs, I have attempted to illustrate that from this focus, a wide range of implications and new questions emerge, ranging way beyond the teacher-student relationship in specific teaching situations. This suggests – and this should not come as a surprise – that a thorough educational response to complexity should be developed holistically, including all dimensions of a school (see, for instance, Bosevska and Kriewaldt 2020) as well as its situatedness in the larger educational system (i.e. as a whole school and whole educational system approach). From this point of view, in fact, this brief exploration of new openings could be further expanded in several directions (e.g. how might we design the spaces of education in such a way that disrupts our understanding of schools as spaces separated from society?), and so we again find ourselves exposed to relationality and complexity. For now, however, I shall end my inquiries by once more foregrounding the pedagogical focus that has been my departure. For, although I have engaged with and shared myriad exemplary teacher experiences, I have purposefully limited myself to a rather generic articulating of helpful perspectives. These perspectives, in fact, only become meaningful if lively engaged with by teachers in their specific teaching contexts. Such a lively engagement, notably, is likely to result in two types of insight: (1) insight of the engaging teacher concerning how to improve his/her teaching, and (2) insight concerning ways in which the helpful perspectives could or should be further improved to become more helpful. It is in this sense, that the outcomes of this book are fundamentally unfinished, and, also, that they would have become richer would I have shaped more diffractive steps with my co-researchers, and, also, with yet other teachers. My curiosity has lately been triggered, for instance, by the following examples of what appear to me powerful attempts of organizing collaborative inquiry within entangled phenomena: (1) educational experiments focused on bringing together students and policymakers into one classroom to together imagine and act toward sustainable urban futures (Hoffman et al. 2021; Oomen et al. 2021), (2) efforts of practicing participatory decision-making within democratic schools (see Korkmaz and Erden 2014), and (3) experiments with university students to undergo and learn from a digital detox (Schnitzler 2017). Bringing helpful perspectives such as those articulated throughout this book in conversation with teachers committed to such exemplary practices – and thus allowing them to be expanded and reshaped – is a crucial path to pursue in moving forward. The embrace of complexity has only just begun.

References

Akkerman, S. F., Bakker, A., & Penuel, W. R. (2021). Relevance of educational research: An ontological conceptualization. *Educational Researcher, 50*(6), 416–424.

Barad, K. (2007). *Meeting the universe halfway: Quantum physics and the entanglement of matter and meaning*. Durham: Duke University Press.

Berry, K. S. (2015). Research as bricolage: Embracing relationality, multiplicity and complexity. In Tobin, K., & Steinberg, S. R. (Ed.), *Doing educational research: A handbook* (2nd ed.) (p. 79–110). Rotterdam: Sense Publishers.

Biesta, G. J. J. (2010). *Good education in an age of measurement: Ethics, politics, democracy*. Boulder, CO: Paradigm Publishers.

Biesta, G. (2016). Improving education through research? From effectiveness, causality and technology to purpose, complexity and culture. *Policy Futures in Education, 14*(2), 194–210.

Bijlsma, H., Filemon, L., Konig, D., Mudde, R., & van Schaik, M. (2017). *De staat van de leraar 2017*. Utrecht: Onderwijs coöperatie.

Bijlsma, H., de Ruig, N., Scheerens, J., & Filemon, L. (2018). *De staat van de leraar 2018*. Utrecht: Onderwijs coöperatie.

Bosevska, J., & Kriewaldt, J. (2020). Fostering a whole-school approach to sustainability: Learning from one school's journey towards sustainable education. *International Research in Geographical and Environmental Education, 29*(1), 55–73.

Crowell, S., & Reid-Marr, D. (2013). *Emergent teaching: A path of creativity, significance, and transformation*. Lanham, MD: Rowman & Littlefield.

De Jong, L. (2021). *Teacher professional learning and collaboration in secondary schools* [PhD thesis, Leiden University]. Leiden University Scholarly Publications. Retrieved from https://scholarlypublications.universiteitleiden.nl/handle/1887/3176646

Doll Jr., W. E. (1993). *A post-modern perspective on curriculum*. New York, NY: Teachers College Press.

Erasmus University Rotterdam (n.d.). *Educatieve Master primair onderwijs*. Retrieved November 15, 2021, from https://www.eur.nl/essb/master/educatieve-master-primair-onderwijs

Fluijt, D. (2018). *Passend onderwijzen met co-teaching: Studie naar de gemeenschappelijke betekenisverlening van co-teaching teams* [PhD thesis, Utrecht University]. Utrecht University Repository. Retrieved from https://dspace.library.uu.nl/handle/1874/369721

Hoffman, J., Pelzer, P., Albert, L., Béneker, T., Hajer, M., & Mangnus A. (2021). A futuring approach to teaching wicked problems. *Journal of Geography in Higher Education*. Advance online publication. https://doi.org/10.1080/03098265.2020.1869923.

Hogeschool Leiden (n.d.). *Spring – Pabo deeltijd*. Retrieved November 15, 2021, from https://www.hsleiden.nl/pabo-deeltijd/spring/index.html

Ingold, T. (2008). Bindings against boundaries: Entanglements of life in an open world. *Environment and Planning A: Economy and Space, 40*(8), 1–15.

Inspectie van het Onderwijs (2019). *De staat van het onderwijs 2019*. Den Haag: Inspectie van het Onderwijs.

Kincheloe, J. L., & Tobin, K. (2015). Doing educational research in a complex world. In Tobin, K., & Steinberg, S. R. (Ed.), *Doing educational research: A handbook* (2nd ed.) (p. 3–13). Rotterdam: Sense Publishers.

Korkmaz, H. E., & Erden, M. (2014). A delphi study: The characteristics of democratic schools. *The Journal of Educational Research, 107*(5), 365–373.

Meirieu, P. (2016). *Pedagogiek: De plicht om weerstand te bieden* (S. Verwer, Vert.). Culemborg: Uitgeverij Phronese.

Oomen, J., Hoffman, J., & Hajer, M. A. (2021). Techniques of futuring: On how imagined futures become socially performative. *European Journal of Social Theory*. Advance online publication. https://doi.org/10.1177/1368431020988826.

Palmer, P. J. (2017). *The courage to teach: Exploring the inner landscape of a teacher's life* (20th anniversary ed.). Hoboken, NJ: Jossey-Bass.

Polanyi, M. (2009). *The tacit dimension*. Chicago, IL: University of Chicago Press.

Schnitzler, H. (2017). *Kleine filosofie van de digitale onthouding*. Amsterdam: De Bezige Bij.

Stolk, E., Boufangacha, Z., & Althuizen, P. (2019). *Ultimatumbrief over PO en VO aan Rutte*. Retrieved 16-11-2021 from https://www.fnv.nl/getmedia/e5244d68-acbf-4466-9e37-fb0cf5703141/Ultimatumbrief-PO-en-VO-aan-Rutte.pdf

Tobin, K., & Steinberg, S. R. (Ed.) (2015). *Doing educational research: A handbook* (2nd ed.). Rotterdam: Sense Publishers.

Milton Keynes UK
Ingram Content Group UK Ltd.
UKHW051004061123
432050UK00004B/40